THE BOTANIST IN
AND ADJACENT IS

An Annotated Check-list of the Vascular Plants of the Islands of Skye, Raasay, Rona, Rum, Eigg, Muck, Canna, Scalpay, and Soay

Third Edition

C.W. Murray

Prabost, Isle of Skye & BSBI Recorder for vice-county 104

and

H.J.B. Birks

University of Bergen, Norway & University College London

with contributions from

the late R.M. Murray, Prabost Weather Station,

C. Jenks, University of Bergen, Norway,

and S.J. Bungard, Raasay

Prabost and Bergen

2005

1st edition 1974 published by Portree High School

2nd edition 1980 published by Botanical Society of the British Isles

This edition privately published by C.W. Murray and H.J.B. Birks with support from the Botanical Society of the British Isles, the Glasgow Natural History Society, and the University of Bergen

Obtainable from H.J.B. Birks, Department of Biology, University of Bergen, N-5007, Bergen, Norway

Cover photo: The Cuillin Hills from The Storr, Isle of Skye.

Photographed by John Birks.

Printed by University College London, England.

ISBN: 0-9548971-0-2

CONTENTS

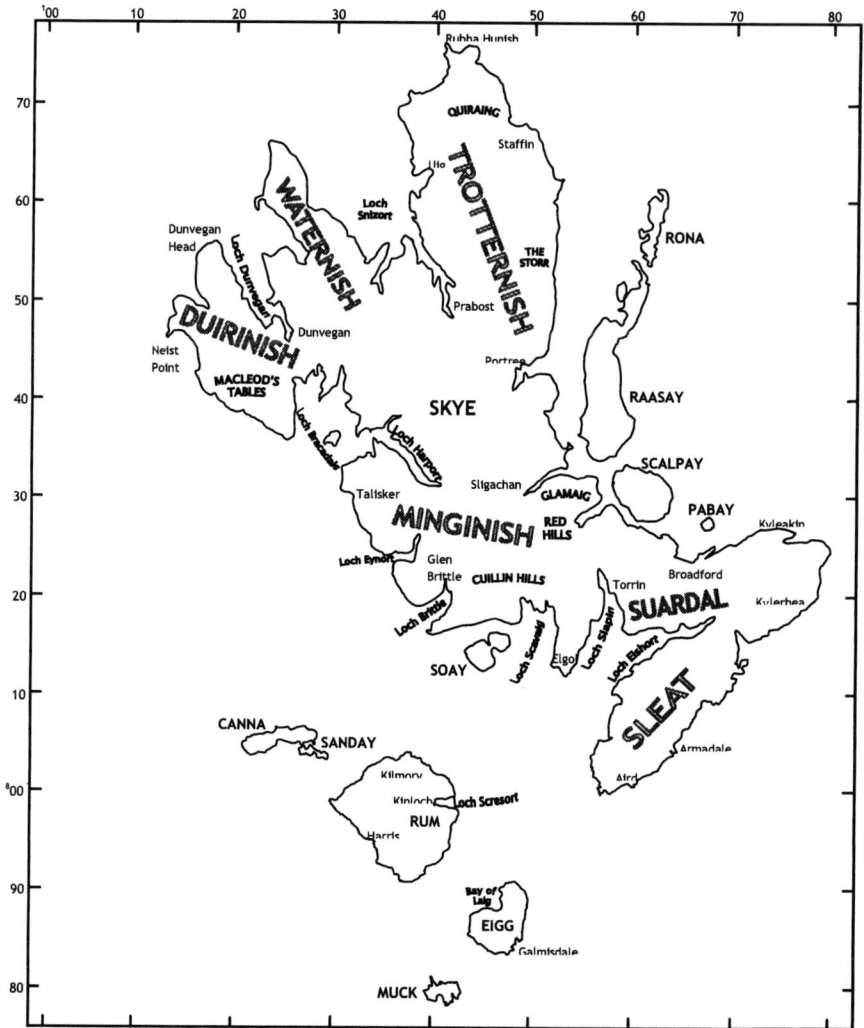

Figure 1. Place names on the Isle of Skye and adjacent islands that form the botanical recording unit of vice-county 104 (North Ebudes)

INTRODUCTION

Introduction

The Isle of Skye, the nearby islands of Raasay, Rona, Rum, Eigg, Muck, Canna, Sanday, Scalpay, and Soay, and other small islands comprise the botanical recording unit of Watsonian vice-county (v.c.) 104 (North Ebudes) (Figure 1). Although Skye is by far and away the largest island in the vice-county, the other islands support many interesting plants, several of which have not, as yet, been found growing on Skye itself.

Recording of plants on Skye goes back before 1700, when Martin Martin (a native of Skye), in "A Description of the Western Islands of Scotland" (1703) included, in a list of plants used in medicine, a quotation from Ray's "Synopsis" (1696)- "**Dryas octopetala** grows on marble in divers parts about Christ-church in Strath; never observed before in Britain and but once in Ireland". Other botanists contributed during the 18th and 19th centuries; the longest lists, after those included in Lightfoot's "Flora Scotica" (1770), coming from M. A. Lawson (1868) and W. R. and E. F. Linton (1884). C. E. Salmon published lists made by A. Wallis in 1910 and 1916, and J. W. Heslop Harrison concentrated in the 1930s on the then little-known islands of the rest of vice-county 104 (Raasay, Eigg, Rum, Soay, etc.)

The BSBI Mapping Scheme of the 1950s showed great blanks in the records for NW Skye, and a Field Meeting based on Dunvegan (1958) proved that interesting records could still be made. This meeting was also responsible for C.W. Murray becoming interested enough to continue plant recording. In 1966, H.J.B. Birks came to Skye to start research on the vegetational history of the island, spending many weeks, over six summers, and along with H.H. Birks, adding many 'firsts'. Also in 1966, C.W. Murray, who has lived on Skye since 1955, was appointed Recorder for vice-county 104 by the Botanical Society of the British Isles. Others have come on holiday and found plants that had escaped the notice of the 'locals'. A manuscript Flora was contributed by F. R. Browning of Tunbridge Wells, and Father N. Dennis prepared a list of the pre-1930 published sources. "Past and Present Vegetation of the Isle of Skye" by H. J. B. Birks was published in 1973, and the following summer the first check-list in "The Botanist in Skye" was published with the help of staff and pupils of Portree High School, after the manuscript had been read by J. C. Brownlie, B. L. Burtt, and J. G. Roger of the Botanical Society of Edinburgh.

In 1980 a much revised and updated check-list "The Botanist in Skye" was published by the Botanical Society of the British Isles. At that time just over 750 species had been recorded (many hybrids and critical subspecies were omitted) and the appendix of 'doubtful' records then contained 67 species. This edition went out-of-print around 1990. As a result of recent botanical explor-ation on Raasay, Eigg, Muck, Canna, Soay, etc. it was decided to expand this new edition to provide an annotated check-list of all the native and introduced species of flowering plant and fern recorded on Skye and the other islands that form the vice-county of North Ebudes (v.c. 104).

Rum has been extensively surveyed by members of the former Nature Conservancy, Nature Conservancy Council, and related bodies, and a check-list was published by Eggeling (1965). Eigg, Muck, and Canna were surveyed by J.C. Brownlie and J.G. Roger in the 1970s. A check-list for Muck was published by R.H. Dobson and R.M. Dobson (1985) and a plant list for Canna was included in Campbell (1984), and updates were published by Birks *et al.* (1991). In the 1990s J. Bevan, P.F. Braithwaite, and C.W. Murray updated the lists for Eigg and Muck while M. Gregory has worked in southern Skye. A survey of freshwater lochs on Skye by Butterfield and Bell (1996) provided several important records for submerged aquatic macrophytes. S.J. Bungard has extensively searched Raasay and Rona (Bungard 1993 and subsequent updates).

Check-list

The check-list follows the order of Stace (1991), and, as far as possible, the plant nomenclature of Stace (1991). Each species entry includes the current scientific name, its scientific name as used in the second edition of the check-list, its popular English and Gaelic names; a brief description of the species' habitat and its relative frequency; its plant-geographical element (a group of species with similar geographical distribution within Europe) according to Preston and Hill (1997) (for details, see below); and its distribution in the vice-county in terms of islands (Skye, Raasay (abbreviated to Ra), Rona (Ro), Rum (Ru), Eigg (E), Muck

(M), Canna (C), Scalpay (Sc), and Soay (So)). Localities known to us are given for species with a limited distribution, whereas for those species where only a grid-square number is given, the exact locations are not known to us, as of June 2004. Gaelic names for native species were contributed by six speakers of Skye Gaelic, the rest follow the list prepared by J.W. Clark and I. MacDonald (1999) for the Botanical Society of the British Isles. Spelling of place names follows the Ordnance Survey 1:50,000 Landranger Series of Great Britain, except for Rum. Sir George Bullough changed the name to Rhum but the true Gaelic spelling is Rum or Rùm (Magnusson, 1997).

Many of the records for islands such as Eigg, Muck, Canna, Scalpay, and Soay are based on our own observations, supplemented from lists such as Braithwaite (2001) and from those provided by Mrs F. Aungier (Canna) and N. Taylor (Muck). Many recent records from Raasay and Rona are the result of detailed surveys by S.J. Bungard. Many of the records for Rum are based on our observations or the check-list by Eggeling (1965), whereas those from Muck are mainly based on Dobson and Dobson (1985), supplemented by later recorders. Most of the smaller islands have not yet been surveyed, though Trodday, Flodigarry Island, and Isay were visited briefly in 1987.

All the 10 km grid squares on Skye, Raasay, Rona, Scalpay, Soay, and Canna lie within the 100 km square 18(NG), whereas all the squares (except 18/30 N. Rum) for Rum, Eigg, and Muck lie within the 100 km square 17(NM). As no ambiguities arise in the 10 km numbers, the prefixes 18(NG) and 17(NM) are omitted throughout. In the species accounts, the 10 km squares from which a species has been recorded are given in numerical order within Skye and Raasay. Square numbers are omitted when a species occurs in all squares. If the species is recorded in 25 or more squares of the 35, then those squares it has **not** yet been found in are listed. Some of the grid squares with very small land areas have been included with adjoining squares; 18/13 with 18/23, 18/31 with 18/41, and the minute fractions of 17/59 and 18/82 with 18/50 and 18/72, respectively. Rum, Eigg, Muck, Scalpay, and Rona are treated here as separate recording units even though they lie within two or more 10 km squares. Soay is recorded separately from Skye square 18/41. The recording units (10 km grid squares or entire islands) used are indicated on Figure 2.

4

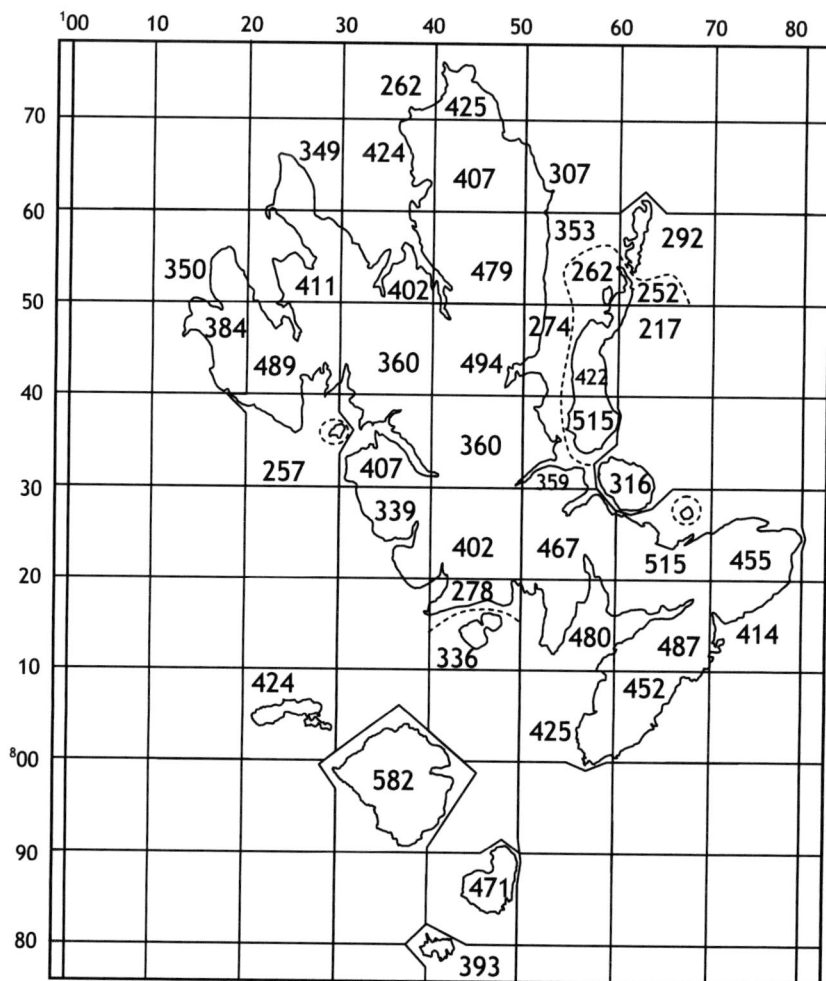

Figure 2. The total number of species of vascular plants (ferns and flowering plants) recorded in each 10 km grid square of the National Grid on the Isles of Skye, Raasay, Rona, Rum, Eigg, Muck, Canna (and Sanday), Scalpay, and Soay to June 2004.

The species marked with '*' after the plant name in the check-list have not been seen by us in v.c. 104 in 40+ years of fieldwork. These species may, of course, have been present in the past or we have overlooked them.

The number of species is now over 800 (some hybrids and critical subspecies are omitted) and the appendix of 'doubtfuls' contains 59 species (see Appendix I). Trees and shrubs that are neither natives nor garden escapes are given in Appendix II on Estate and Forest Enterprise Woodland. The Forest Enterprise list also includes 'amenity planting' of hardwoods and shrubs, as well as the better-known stands of conifers.

Appendix III lists the botanists who have contributed the majority of the plant records for Skye and the rest of v.c. 104.

Plant geographical elements

"Skye is remarkable amongst the Hebridean islands for its variety of geology, topography, and climate, which creates a wide range of habitats for plants and animals. As a result, both the flowering plant and cryptogamic floras of Skye are very rich, and contain members with contrasting geographical and ecological affinities." (Birks and Birks, 1974)

In this section we provide a new survey of the plant geography of the vascular plants (flowering plants and ferns) of v.c. 104 that totally supports this statement.

Almost every species occurring in v.c. 104 has been allocated to one of the floristic elements defined by Preston and Hill (1997). These elements are based on two simple criteria – (i) species present as natives in one or more of the major terrestrial biomes (Arctic, Boreal, Temperate, Southern) that show a strong north-south distribution within Europe (Table 1), and (ii) their eastern limits (Oceanic, Suboceanic, European, Eurosiberian, Eurasian, Circumpolar) (Table 2). Each species is allocated to a major biome category and an eastern limit category and the floristic element represents a combination of these two criteria, e.g. the Oceanic Temperate element consists of species in the Oceanic eastern limit category and the Temperate major biome category. In addition, there are three special elements (Table 3) for species with predominantly Mediterranean distributions (Mediterranean-Atlantic, Submediterranean-Subatlantic, and Mediterranean-Montane) and for species confined to Britain and

6

Ireland (Endemic). The total numbers of species occurring on Skye and adjacent islands in v.c. 104 and in Britain and Ireland in each floristic element are given in Table 4. This analysis shows how geographically diverse the flora of Skye and adjacent islands is with species belonging to 42 of the 44 geographical elements represented in Britain and Ireland as a whole.

Table 1. Classification of species into major biome categories.

Major biome category	Abbreviation used	Main distribution
Arctic-Montane	ArMo	North of the tree-line or on mountains above the tree-line or both
Boreo-Arctic Montane	BAMo	In both Arctic-Montane and Boreal-Montane zones
Wide-Boreal	WiBo	In Arctic-Montane, Boreal-Montane, and Temperate zones
Boreal-Montane	BoMo	In both Boreal-Montane and Temperate zones
Wide Temperate	WiTe	In Boreal-Montane, Temperate, and Southern-Temperate zones
Temperate	Temp	In the cool-temperate, broad-leaved deciduous forest zone
Southern-Temperate	SoTe	In both Temperate and the Southern (Mediterranean) zones
Southern	Sout	In the warm-temperate zone south of the broad-leaved deciduous forest zone which in Europe is represented by the Mediterranean zone

Table 2. Classification of species into eastern limit categories.

Eastern limit category	Abbreviation used	Main distribution
Oceanic	Ocea	Confined to western Europe (Norway, W. Denmark, Low Countries, Britain, Ireland, W. France, and the Atlantic fringe of Iberia)
Suboceanic	SubO	Confined to western and central Europe (occurring west of a line from the Baltic to the Adriatic)
European	Euro	Widespread in Europe but with an eastern limit west of 60ºE
Eurosiberian	ESib	Widespread in Europe and western Asia, with an eastern limit between 60ºE and 120ºE
Eurasian	EAsi	Widespread in Europe and Asia, with an eastern limit of 120ºE
Circumpolar	Circ	Present in Europe, widespread in Asia and also present in North America

Table 3. Additional floristic elements.

Element	Abbreviation used	Main distribution
Mediterranean-Atlantic	Med Atl	Mediterranean zone and the Atlantic fringe of Europe. Absent from Central Europe
Submediterranean-Subatlantic	SubM SubA	Broader than Med Atl and often extends into the south-western parts of central Europe
Mediterranean-Montane	Med Mont	Mainly in the mountains of the Mediterranean zone but in habitats too warm to be considered Arctic or Boreal
Endemic	Endm	Confined to Britain and Ireland

Table 4. Numbers of species in the Preston and Hill (1997) floristic elements on Skye and adjacent islands (v.c. 104) and in Britain and Ireland. Numbers in brackets indicate planted species in v.c. 104.

Element	Abbreviation	v.c. 104	Britain and Ireland
European Arctic-Montane	Euro ArMo	14	29
Eurosiberian Arctic-Montane	ESib ArMo	3	6
Eurasian Arctic-Montane	EAsi ArMo	2	3
Circumpolar Arctic-Montane	Circ ArMo	19	41
Oceanic Boreo-Arctic Montane	Ocea BAMo	1	1
European Boreo-Arctic Montane	Euro BAMo	4	10
Eurosiberian Boreo-Arctic Montane	ESib BAMo	1	2
Circumpolar Boreo-Arctic Montane	Circ BAMo	13	25
Eurosiberian Wide-Boreal	ESib WiBo	1	1
Eurasian Wide-Boreal	EAsi WiBo	1	1
Circumpolar Wide-Boreal	Circ WiBo	13	17
Oceanic Boreal-Montane	Ocea BoMo	4	7
Suboceanic Boreal-Montane	SubO BoMo	2	5
European Boreal-Montane	Euro BoMo	13	27
Eurosiberian Boreal-Montane	ESib BoMo	5	9
Eurasian Boreal-Montane	EAsi BoMo	1 (1)	5
Circumpolar Boreal-Montane	Circ BoMo	24	50
Oceanic Boreo-Temperate	Ocea BoTe	5 (1)	8
Suboceanic Boreo-Temperate	SubO BoTe	7	8
European Boreo-Temperate	Euro BoTe	30 (1)	48
Eurosiberian Boreo-Temperate	ESib BoTe	47 (1)	69

Eurasian Boreo-Temperate	EAsi BoTe	28	38
Circumpolar Boreo-Temperate	Circ BoTe	47	64
Oceanic Wide-Temperate	Ocea WiTe	0	1
European Wide-Temperate	Euro WiTe	2	3
Eurosiberian Wide-Temperate	ESib WiTe	11	11
Eurasian Wide-Temperate	EAsi WiTe	5	5
Circumpolar Wide-Temperate	Circ WiTe	12	14
Oceanic Temperate	Ocea Temp	20	48
Suboceanic Temperate	SubO Temp	19 (1)	28
European Temperate	Euro Temp	109 (8)	297
Eurosiberian Temperate	ESib Temp	29 (3)	120
Eurasian Temperate	EAsi Temp	11 (2)	38
Circumpolar Temperate	Circ Temp	10	26
Oceanic Southern-Temperate	Ocea SoTe	4	25
Suboceanic Southern-Temperate	SubO SoTe	13	54
European Southern-Temperate	Euro SoTe	38 (1)	106
Eurosiberian Southern-Temperate	ESib SoTe	26(2)	81
Eurasian Southern-Temperate	EAsi SoTe	6	17
Circumpolar Southern-Temperate	Circ SoTe	6	13
Mediterranean-Atlantic	Med Atl	4	69
Submediterranean-Subatlantic	SubM SubA	6 (1)	47
Mediterranean-Montane	Med Mont	0	6
Endemic	Endm	2	48
Unassigned	-	2	10

	J	F	M	A	M	J	J	A	S	O	N	D	Year
Sunshine													
Average (hrs)	38	74	101	146	181	167	124	125	97	74	42	30	**1199**
Possible (hrs)	188	235	336	399	477	495	496	442	357	290	204	163	**4082**
Average as % of possible	20	31	30	36	38	34	24	28	27	25	21	18	**28**
Sunniest (top of range) (hrs)	83	121	139	257	300	250	199	272	144	113	78	66	**1448**
Dullest (bottom of range) (hrs)	8	36	49	87	113	95	56	65	50	41	25	11	**1025**
Temperature													
Average (°C)	4	4	5	7	10	12	13	13	11	9	6	4	**8**
Average monthly max (°C)	10	10	12	15	20	22	21	22	18	16	12	11	**16**
Average monthly min (°C)	-4	-4	-3	-2	1	4	6	5	3	2	-2	-3	**3**
Extremes: Max (°C)	13	13	16	21	26	27	27	28	27	21	14	14	**28**
Min (°C)	-8	-9	-6	-4	-2	1	3	3	0	-1	-6	-8	**-9**
Rainfall													
Average (mm)	174	116	147	86	87	102	114	127	190	209	215	204	**1771**
Top of range (monthly totals) (mm)	372	337	369	178	217	192	228	317	364	377	411	366	**2236**
Bottom of range (monthly total) (mm)	37	9	28	4	15	19	45	20	51	40	82	60	**1372**
Wintry Days													
Snow fall	7	5	6	3	1	0	0	0	0	0	3	5	**30**
Snow lying	7	6	3	1	0	0	0	0	0	0	2	4	**23**
Air frost	9	10	6	3	1	0	0	0	0	0	4	7	**40**
Ground frost	20	19	18	14	7	2	1	1	2	6	14	17	**121**
Wind													
Gales (days)	8	5	6	2	2	1	1	1	4	5	6	8	**49**
Mean speed (mph)	16	14	15	13	12	12	11	11	13	15	14	15	**13**
N or NW (% 09 GMT obs)	13	9	15	21	22	22	24	23	17	14	16	14	**18**
W or SW (% 09 GMT obs)	31	23	31	30	26	37	34	36	38	35	34	32	**32**
S or SE (% 09 GMT obs)	41	46	36	27	30	26	22	30	32	38	35	40	**34**
E or NE (% 09 GMT obs)	16	19	17	22	22	14	11	12	14	12	15	15	**16**

Table 5. CLIMATE IN SKYE. Summary of 30 years observations (1959-89), Meteorological Office climate station at Prabost (200 ft/61 m)

CLIMATE, GEOLOGY, AND LANDFORMS IN SKYE

Geology and relief, together with topographical names, are summarised in Figures 1, 3, and 4, the maps being referenced to each other and to the check-list by the National Grid Square numbers.

Climate averages can be very misleading, especially in Skye, where weather is constantly changing from hour to hour, and from day to day. Any particular month can be quite different from one year to the next, though there is a marked tendency for February-June to be sunnier and drier than the rest of the year (Table 5). To show this variation, the upper and lower limits of a 30-year range (1959-89) at the Meteorological Office climate station at Prabost have been given as well as the averages, and absolute extremes of temperature. While these figures can be taken as representative of lowland Skye, there is likely to be a wide variation in a mountainous island so deeply dissected by the sea. With an average lapse rate of 6°C/1000 metres the highest point in Skye, Sgurr Alasdair in the Cuillin, will be about 6°C colder than Prabost, but the surprisingly high frost figures at Prabost suggest that large variations of temperature can occur near sea level. Rainfall normally increases with altitude, and a monthly rain gauge on the Storr, 472 metres above Prabost, averaged 2108 mm compared with 1702 mm at the lower station, over the seven years 1962-68. Another at the head of Loch Slapin (1969-75) averaged 2933 mm, while Prabost averaged 1616 mm in the same period. The Meteorological Office rainfall map, based on annual averages 1941-70 gives 3400 mm over the Cuillin and just under 1400 mm at the seaward ends of the northern peninsulas, the driest parts of Skye.

Geology also varies even more than is shown on the geological map (Figure 3). As well as the large area in north and west Skye, basalt occurs, along with the related but more coarsely crystalline dolerite, in numerous sills and dykes in the underlying Jurassic of north Skye and the Elgol peninsula, and in the gabbro of the Cuillin and Blaven. The Jurassic alternates between sandstone, limestone, and shale, the plastic character of the shale resulting in the massive landslip zone in front of the Storr and Quiraing scarp. The basalt also alternates between tough fine-grained grey rock, much darker ultra-basic layers, and softer slaggy material, producing the characteristic terraced landscape. Away from cliffs and the upper slopes, the basalt, and the Jurassic as well, are often obscured by acid peat, contrasting botanically with these basic rocks. Acid rocks do occur, as in the granite of the Red Hills, and the Torridon sandstone which alternates with the Cambrian limestone, botanically the most productive rock in Skye. The limestone is usually altered to marble, as at Suardal and Ord.

Marked contrasts in landscape occur, as between typical Lewisian landscape in Rona and N. Raasay and the gentler Sleat peninsula. The alpine type landscape of the Cuillin differs markedly from the much smoother Red Hills, as does the spectacular Storr-Quiraing escarpment from the extensive moorland between it and the Cuillin.

THE VEGETATION OF SKYE

Skye is divisible into six major regions on the basis of the underlying bedrock geology and topography (Figures 3 and 4). These regions are Sleat (low-lying, generally acid Lewisian gneiss, Torridonian sandstone, and Moine schist with some Cambrian limestone at Ord), Kyleakin (low- and high-altitude Torridonian sandstone), Red Hills (granite), Cuillin Hills (mainly gabbro), Suardal (mainly Cambrian limestone at low altitudes), and Northern Skye (Tertiary basalt and dolerite overlying Jurassic limestone and shale). These regions support a wide range of vegetation types or plant communities, as a result of the many different types of soils, landforms, altitudes, and land-use patterns within each region (Birks, 1973). Each region is of considerable interest to the botanist, not only for the flora it supports, but also for the intrinsic interest in understanding how and why so many contrasting plant communities grow together within the comparatively small area of the Isle of Skye. No other area of comparable size in Western Europe has such a wide range of plant communities. Rather than discussing each community individually, the major vegetation types that a visitor is likely to encounter in each of the regions are described in general terms. Although botanists visiting Skye tend to concentrate on the well-known areas such as Suardal and Trotternish, and to ignore regions such as the Red Hills, Sleat, and Kyleakin, each region abounds in interest for the enthusiastic, energetic, and enquiring field botanist.

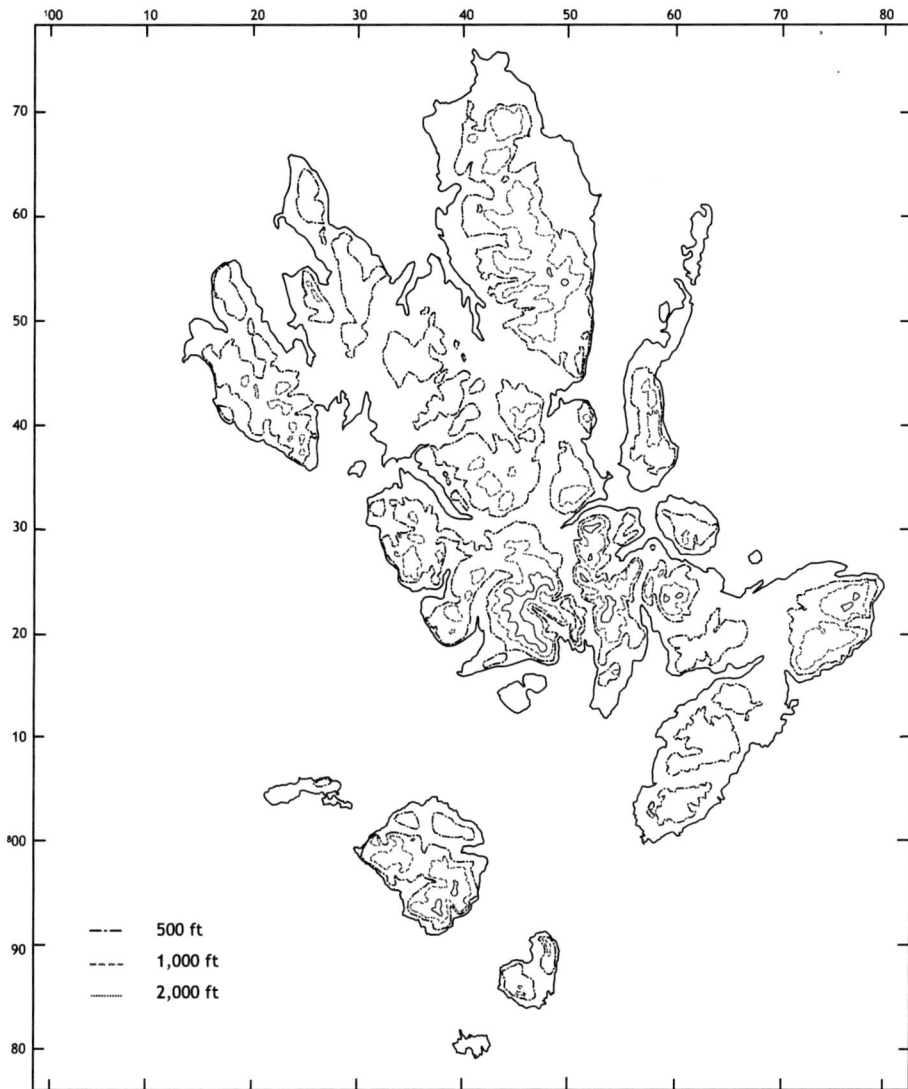

Figure 3. Relief on the Isle of Skye and adjacent islands.

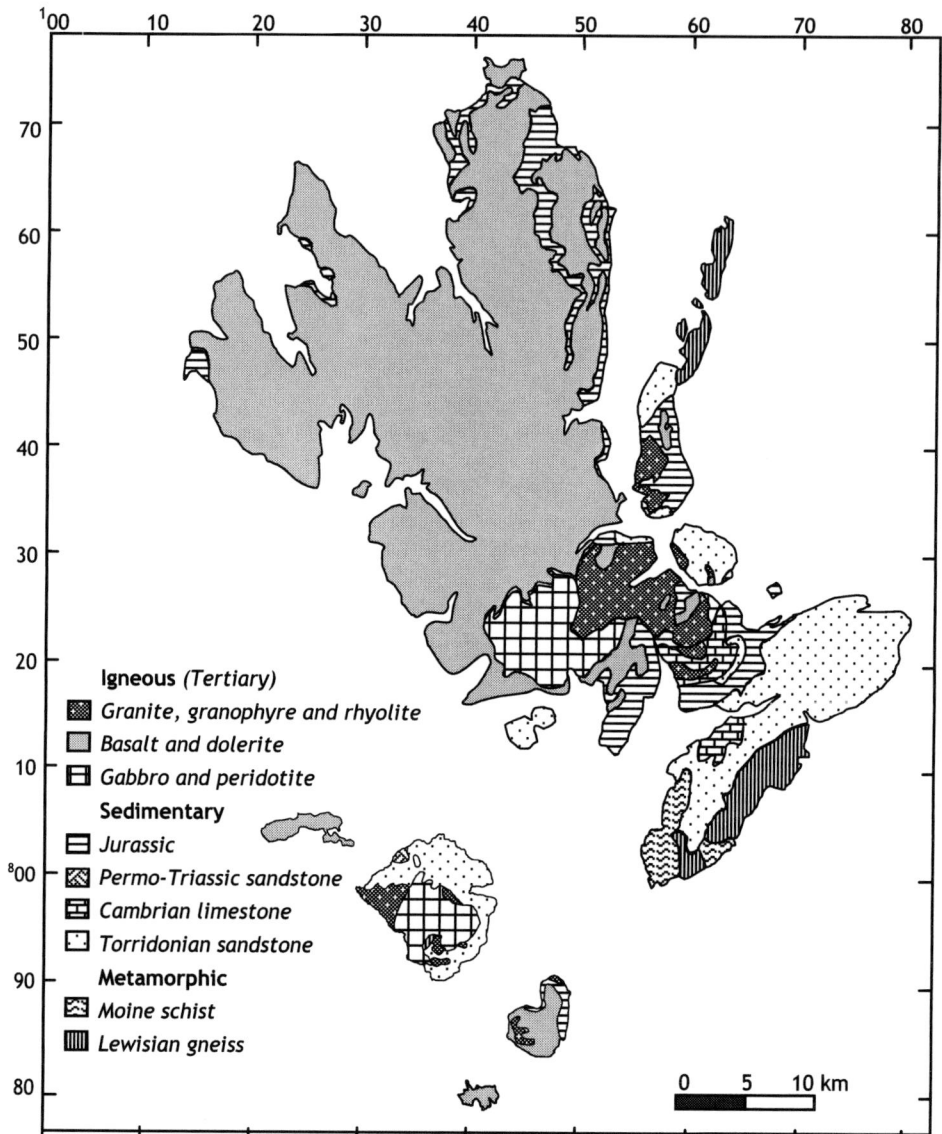

Figure 4. Bedrock geology of the Isle of Skye and adjacent islands

The acid rocks, the low-lying and generally rather subdued topography, and the high rainfall of Sleat result in the soils being waterlogged and peaty. Bog vegetation is thus the commonest vegetation type in this region. *Scirpus cespitosus-Calluna vulgaris* bog, commonly with *Potentilla erecta, Erica tetralix*, and some *Sphagnum* occurs extensively on shallow (2 m or less) peat throughout Sleat. It has usually been cut for fuel. In less disturbed and wetter areas, usually some distance from habitations and where the peat is anything up to 6 m deep, *Scirpus-Eriophorum angustifolium* bog is found with an abundance of *Sphagnum* hummocks and lawns, through which grow *Myrica gale, Erica tetralix, Calluna vulgaris, Narthecium ossifragum, Molinia caerulea, Rhynchospora alba* and the red, worm-like liverwort *Pleurozia purpurea*. Plants of interest in these bogs include the insectivorous *Drosera anglica*, the slender sedge *Carex pauciflora*, and the critical *Pedicularis sylvatica* ssp. *hibernica*. In some areas there are large bog pools with scattered plants of *Menyanthes trifoliata* and *Utricularia* spp. in the deepest water, and with *Carex limosa* at the edges. Along streams and rills draining through these boggy areas, dense tussocks of *Molinia caerulea* occur with well-grown bushes of *Myrica gale*. Although these two dominants leave little room for other species, *Carex echinata, Pinguicula vulgaris, Anagallis tenella*, and *Drosera intermedia* add to the interest of this community. The rare *Rhynchospora fusca* occurs in similar vegetation in Glen Sligachan.

On better drained sites, such as rock knolls and hillsides where bog has not developed, the major communities are either *Calluna vulgaris* heather-moor with *Erica cinerea, Agrostis capillaris*, and *Carex binervis*, or *Agrostis-Festuca* grassland with *A. capillaris, A. canina, F. ovina, Anthoxanthum odoratum, Galium saxatile*, and *Potentilla erecta*. This grassland is the principal sheep pasture on Skye, and there is good evidence to indicate that it is derived, as a result of extensive grazing and burning, from *Calluna* moor.

Sleat and Kyleakin are more sheltered from Atlantic storms than elsewhere on Skye, with the result that woodland is commonest in these regions. Analyses of fossil pollen preserved in bog and loch sediments on Skye indicate that since the last ice age, woodlands were once widespread in Sleat and Kyleakin but that as a result of clearance by man and of the extensive development of bog in the last 4000 years, woodland has become restricted there to steep rocky slopes. Pollen analyses from sites in Northern Skye and on the lower slopes of the Cuillin Hills suggest that woodland was never common in these areas, presumably as a result of the high frequency of wind in these regions.

The most extensive woods in Sleat and Kyleakin occur east of Kyleakin, near Loch na Dal, near Kylerhea, near Ord, and near Aird. They all occur on slopes that are too steep and too rocky for either bog development or widespread clearance. They are dominated by *Betula pubescens* and *Corylus avellana*, with some *Sorbus aucuparia, Ilex aquifolium,* and *Quercus petraea. Alnus glutinosa* and *Salix* spp. occur in damp sites, whereas *Fraxinus excelsior* favours the richest soils associated mainly with the Cambrian limestone at Ord.

On the poorest, most acid soils birch, oak, rowan, and holly predominate with a field layer dominated by *Vaccinium myrtillus, Deschampsia flexuosa, Galium saxatile,* and *Blechnum spicant.* Mosses and liverworts abound in these woods, particularly on north- or east-facing boulder-strewn slopes. The filmy fern *Hymenophyllum wilsonii* is frequent in these situations. Ungrazed areas of acid birchwoods are characterized by an abundance of *Luzula sylvatica* and tall ferns such as *Dryopteris affinis, D. aemula,* and *Gymnocarpium dryopteris.* On richer soils, hazel tends to dominate, with a species-rich field layer of *Primula vulgaris, Oxalis acetosella, Viola riviniana, Circaea intermedia, Lysimachia nemorum,* and *Anthoxanthum odoratum.* Ferns, mosses, and liverworts are again frequent. Lichens are particularly conspicuous, especially on the larger hazel trees. The richest soils, commonly associated with the Cambrian limestone, support ash and hazel, with *Brachypodium sylvaticum, Primula vulgaris*, and *Filipendula ulmaria* in the field layer. Species of interest in these hazel and ash woods include *Allium ursinum, Listera ovata, Viburnum opulus, Prunus padus, Melica nutans, Cephalanthera longifolia* and *Paris quadrifolia.* In areas protected, in some way, from grazing, such as large inaccessible ledges in basic ravines, the field layer changes dramatically, with an abundance of 'tall-herbs' such as *Cirsium heterophyllum, Filipendula ulmaria, Geum rivale, Crepis paludosa, Trollius europaeus,* and *Valeriana officinalis.* It is likely that these beautiful communities were once commoner in woods on Skye before the introduction of sheep.

A feature of considerable international importance in these woods is their remarkable bryophyte and fern flora. Besides the abundant woodland mosses and liverworts that occur on the boulders, trees, rotten logs, and ground in these woods, small ravines, gullies, and cascades provide shaded, humid habitats for many moisture-demanding plants. These include *Hymenophyllum tunbrigense, Dryopteris aemula*, and the inconspicuous algal-like gametophyte of *Trichomanes speciosum*, and several rare mosses and liverworts whose world distributions are Atlantic and/or Tropical, with occurrences in places as far afield as Iberia, Macaronesia, tropical Africa and South America, and the West Indies. It is of considerable interest and importance to botanists and conservationists that many of these species grow at or near their northernmost known world localities in the woods of Sleat and Kyleakin.

The vegetation of the Kyleakin area, outside the woods, is, in the lowlands, not dissimilar to that in Sleat, with widespread bog, heather moor, and grassland. The ground above about 500 m supports a range of montane communities. On flat areas between 300 and 500 m an upland bog-type characterized by abundant *Eriophorum vaginatum, Calluna vulgaris*, and *Empetrum nigrum* ssp. *nigrum* occurs commonly. This community has affinities with the widespread blanket bog of the Eastern Highlands and the Pennines, but the Skye community curiously lacks the characteristic *Rubus chamaemorus*.

With increasing altitude in the Kyleakin Hills the widespread *Calluna* moors of the lower slopes are replaced by a dwarf, wind-pruned mat of *Calluna vulgaris* interspaced by conspicuous carpets of the hairy moss *Racomitrium lanuginosum*, through which grow *Festuca vivipara, Antennaria dioica*, and *Salix herbacea*. Above 650 m *Calluna* disappears and the major vegetation of the summit ground is *Carex bigelowii- Racomitrium lanuginosum* heath with *Alchemilla alpina, Vaccinium myrtillus, Galium saxatile, Salix herbacea, Sibbaldia procumbens, Festuca vivipara*, and, more rarely, *Arctostaphylos alpinum, Loiseleuria procumbens, Gnaphalium supinum*, and *Diphasiastrum alpinum*. This community is dominated by an almost continuous mat of *Racomitrium lanuginosum*, the shoots of which invariably point one way, namely away from the direction of the prevailing wind. In the most exposed sites open *Juncus trifidus-Festuca ovina* 'fell fields' occur with *Luzula spicata* and diminutive plants of *Salix herbacea*. In areas sheltered from the wind and where some snow lies late, straw-coloured tussocks of *Nardus stricta* are conspicuous, along with abundant *Vaccinium myrtillus*, *Huperzia selago*, and *Diphasiastrum alpinum*. Cliffs and screes are relatively rare in the Kyleakin Hills but those that do occur support mats of *Racomitrium lanuginosum* with abundant *Empetrum nigrum* ssp. *hermaphroditum* and *Carex bigelowii*, and the rare but elegant mountain hawkweed *Hieracium holosericeum*.

The vegetation of the granitic Red Hills is fairly similar to the vegetation of the Kyleakin Hills. The major difference results from the greater extent of cliff and stable block-scree in the Red Hills. These screes, particularly in the shaded north- and east-facing corries of Beinn na Caillich and Beinn Dearg support a community dominated by tall *Calluna vulgaris* and *Vaccinium myrtillus* bushes, amidst which grow in crevices and caves between boulders the ferns *Blechnum spicant* and *Hymenophyllum wilsonii* and an abundance of mosses and liverworts. Several of these bryophytes have remarkable world distributions, with localities not only in Western Britain and Ireland but also in Western Norway, Himalaya, Yunnan, Bhutan, Hawaii, and British Columbia. These damp *Vaccinium-Calluna* heaths provide the main habitat for the diminutive orchid *Listera cordata*. These heaths are best developed in the Red Hills, although they were probably once more widespread on Skye prior to extensive burning and grazing. A second feature of interest in the Red Hills is the local abundance of dwarf-juniper heath above 500 m altitude. These heaths are very rare on Skye, but are best represented on wind-exposed cols and the lower summits of the Red Hills. The frequent occurrence of *Cryptogramma crispa* in the screes is also of interest.

The gabbro of the Cuillin Hills provides spectacular scenery, with 23 serrated peaks over 920 m altitude, narrow summit ridges, deeply cut corries, and massive cliffs and screes. The lower slopes of the Cuillin are covered by vast areas of dull *Molinia caerulea-Calluna vulgaris* bog characterized by abundant, large *Molinia* tussocks. The growth of this grass is encouraged by the seepage of mildly basic water through the peat. Little of botanical interest grows in these bogs, with the exception of the elusive bog orchid *Hammarbya paludosa*, the insectivorous

Pinguicula lusitanica, and the British endemic moss *Campylopus shawii*. Small areas of open gravelly flushes occur wherever there is constant movement of base-rich water from small springs. These flushes support abundant *Schoenus nigricans* where the rate of water flow is slow, or *Eriophorum latifolium*, *Carex hostiana*, *C. dioica*, *C. viridula* ssp. *brachyrrhyncha*, *Eleocharis quinqueflora*, and *Saxifraga aizoides* where the water flow is rapid.

Above about 400 m altitude, the well-drained lower slopes of the Cuillin support either *Agrostis-Festuca* grasslands, often with *Alchemilla alpina* and other montane herbs such as *Silene acaulis* and *Persicaria vivipara*, or species-poor *Calluna* heaths. On the steepest slopes there are extensive screes with little or no vegetation. The extensive gabbro cliffs are comparatively poor botanically. Despite the basic character of the rock its great resistance to weathering does not produce particularly basic soils. The main vegetation on the cliff ledges is similar to that in the Red and Kyleakin Hills, namely *Empetrum nigrum* ssp. *hermaphroditum-Racomitrium lanuginosum* communities with *Alchemilla alpina*, *Antennaria dioica*, and *Solidago virgaurea*. Crevices, earthy screes, flushes, and damp gullies on the higher gabbro cliffs support a sparse but interesting flora, with *Arabis petraea*, *Oxyria digyna*, *Rubus saxatilis*, *Sedum rosea*, *Saxifraga stellaris*, *Silene acaulis*, and, more rarely, *Saussurea alpina*, *Saxifraga oppositifolia*, *S. nivalis*, *Poa alpina*, *Cerastium arcticum*, *Equisetum variegatum*, *Draba norvegica*, and, in its only known British localities, *Arabis alpina*.

The sparse vegetational cover of the gabbro cliffs contrasts with the luxuriant vegetation on the cliffs of Jurassic limestone in Coire Uaigneach on Blaven. There rich, ungrazed, 'tall-herb' communities with *Sedum rosea*, *Alchemilla glabra*, *Trollius europaeus*, and *Saussurea alpina* luxuriate. On the near vertical, damp limestone faces, *Saxifraga aizoides* forms almost pure banks with *S. oppositifolia*, *Thalictrum alpinum*, and the elegant salmon-pink coloured moss *Orthothecium rufescens*. In shaded crevices in the cliffs ferns such as *Asplenium viride*, *Cystopteris fragilis*, and *Polystichum lonchitis* grow with numerous rare mosses and liverworts.

The summits of all the Cuillin Hills are largely devoid of vegetation. What there is, is open *Juncus trifidus-Festuca ovina* 'fell-field' with *Luzula spicata*, *Deschampsia cespitosa* ssp. *alpina*,

Alchemilla alpina, Loiseleuria procumbens, and *Salix herbacea.* Small areas of *Nardus stricta*-dominated snow-patches occur in some of the gullies in the north-facing corries.

The Suardal area, just to the south of Broadford, is dominated by Ben Suardal (310 m). Although of low relief, the area is of considerable botanical interest. The special feature is the extensive outcrop of Cambrian limestone that occurs either as small cliffs or as large areas of fissured limestone pavement with well-developed clint-and-grike structure, such as one sees in the famous Carboniferous limestone areas of the Craven Pennines in Yorkshire or The Burren in County Clare in Ireland. In Suardal there is a great abundance of the arctic-alpine *Dryas octopetala,* growing, as it does at Kishorn, Inchnadamph, Durness, and The Burren, right down to sea level. Besides the abundance of *Dryas,* these heaths are characterized by *Carex flacca, Festuca vivipara, Pilosella officinarum,* and *Thymus polytrichus.* Areas of woodland growing on the limestone are dominated by ash and hazel, and are similar to the ash-hazel woods of Sleat. Treeless areas between the limestone pavements support species-rich *Agrostis-Festuca* grassland with *Anthoxanthum odoratum, Carex pulicaris, Lotus corniculatus, Prunella vulgaris,* and, more rarely, *Botrychium lunaria, Galium boreale, Gentianella amarella,* and *Persicaria vivipara. Potentilla crantzii* and *Carex rupestris* occur in very open turf and limestone ledges. The Suardal grasslands curiously lack *C. capillaris* that is locally common at Kishorn, Inchnadamph, and Durness.

The larger areas of pavement are predominantly treeless, and support a diverse vegetation. Small ferns such as *Asplenium ruta-muraria, A. trichomanes, A. viride,* and *Cystopteris fragilis* grow in crevices in the rock faces, whereas woodland herbs and ferns such as *Sanicula europaea, Epipactis atrorubens, Listera ovata, Paris quadrifolia, Phyllitis scolopendrium,* and *Polystichum aculeatum* grow in the shade and shelter of the deeper grikes. Grassland herbs grow on the small pockets of soil that develop on the clints. Numerous springs arise from within the limestone and these result in several open gravel flushes with an abundance of *Schoenus nigricans, Eriophorum latifolium,* and *Saxifraga aizoides.*

Northern Skye consists of the extensive Tertiary basalt areas and the underlying Jurassic rocks. The vegetation in the lowlands consists of *Scirpus-Calluna* or *Scirpus-Eriophorum*

21

angustifolium bog in waterlogged areas, similar to those in Sleat. *Calluna* heather-moor or *Agrostis-Festuca* grassland favour steep, well-drained sites, whereas hay-meadows characterize the crofting areas of Northern Skye. The commonest enclosed hay-meadow is dominated by *Cynosurus cristatus, Festuca rubra*, and *Poa pratensis*, with a variety of herbs including *Centaurea nigra, Ranunculus repens, Trifolium pratense*, and *T. repens. Platanthera chlorantha* is an attractive but rare member of these meadows. In damper sites that are mown infrequently and grazed rarely by sheep and cattle, *Juncus acutiflorus* meadows predominate with *Filipendula ulmaria, Lychnis flos-cuculi, Achillea ptarmica*, and *Senecio aquaticus*. Hay meadows are declining because of changes in land-use practices.

The lower slopes of the massive and spectacular Trotternish ridge support damp grasslands and flush bogs, with an abundance of *Juncus effusus, Carex echinata*, and *Sphagnum* spp. On steeper ground *Alchemilla*-rich *Agrostis-Festuca* grasslands are widespread. These are heavily grazed by sheep and rabbits. Besides an abundance of *Alchemilla alpina*, several 'microspecies' occur including *A. filicaulis, A. xanthochlora, A. wichurae*, and *A. filicaulis* ssp. *vestita. Ranunculus acris, Rumex acetosa, Botrychium lunaria*, and *Selaginella selaginoides* are also frequent. Towards the basalt cliffs, montane herbs are found in the grazed sward, including *Minuartia sedoides, Sibbaldia procumbens, Silene acaulis*, and *Thalictrum alpinum*. With over-grazing, these grasslands develop into open, eroding basalt screes that are extremely unstable and support scattered plants of *Sagina subulata, Arabis petraea*, and *Poa glauca. Koenigia islandica* can occasionally be found in some of the damper screes.

The vegetation of the basalt cliffs at, for example, The Storr, The Quiraing, Beinn Edra, and Sgurr Mor is similar to that of Coire Uaigneach, with well-developed ledges of 'tall-herbs', crevices rich in ferns, banks of *Saxifraga aizoides*, and shaded gullies and cliffs with a variety of rare species including *Poa alpina, Woodsia alpina, Salix myrsinites* and *Saxifraga nivalis*. In addition there are dry, sun-exposed south-facing cliffs with *Draba incana, Arabis petraea*, and *Poa glauca*.

The summit vegetation of Trotternish is very varied, with large areas of *Carex bigelowii-Racomitrium lanuginosum* heath, *Nardus stricta* grassland, *Juncus squarrosus-Nardus* bogs, and open 'fell-fields'. Springs of acid or mildly basic water are frequent, with dense tufts of the mosses *Philonotis fontana* and *Dicranella palustris*, within which grow *Saxifraga stellaris, Montia fontana, Epilobium alsinifolium,* and *E. anagallidifolium.* Open areas of gravel below these springs on the Storr and Beinn Edra support *Juncus biglumis, J.triglumis, Carex viridula* ssp. *oedocarpa, Deschampsia cespitosa* ssp. *alpina,* and the arctic-subarctic *Koenigia islandica. Cerastium arcticum* grows, in small quantity, in an open, windblasted area below the Storr Summit.

A feature of much of Northern Skye is the spectacular coastline, with several massive sea-cliffs, such as Dunvegan Head, Rubha Hunish, and Waterstein Head. A coastal facies of the inland *Calluna vulgaris* heath commonly occurs near the cliff-edge, with *Danthonia decumbens, Lotus corniculatus, Plantago maritima, Lathyrus montanus,* and *Salix repens* var. *argentea.* The upper ledges and crevices of these sea cliffs support several montane species, including *Saxifraga oppositifolia, S. hypnoides, Sedum rosea, Draba incana,* and *Silene acaulis.* The influence of sea spray becomes increasingly important at lower levels, resulting in vegetation dominated by *Armeria maritima,* with *Plantago maritima, Festuca rubra, Sedum anglicum,* and *Ligusticum scoticum.* Sheltered caves and recesses support the fern *Asplenium marinum.* Ledges below nesting sites of sea birds carry luxuriant growths of *Athyrium filix-femina, Silene dioica,* and *Urtica dioica.*

Deeply cut ravines, often with spectacular waterfalls, provide sheltered, ungrazed habitats in Northern Skye. In these, lush stands of 'tall herbs' occur, with *Luzula sylvatica, Filipendula ulmaria, Cirsium heterophyllum, Geum rivale, Valeriana officinalis,* and *Crepis paludosa.* Fine examples of this vegetation can be seen in the Geary ravine in Waternish, by the Lealt waterfall in Trotternish, and around Lovaig Bay.

The Jurassic limestone cliffs of the east coast of Northern Skye support dense hazel scrub, with an abundance of *Equisetum telmateia, Eupatorium cannabinum, Silene dioica,* and *Valeriana officinalis.* Steep rock outcrops and cliffs harbour an interesting flora, including *Arabis hirsuta, Geranium lucidum, Torilis japonica,* and *Trifolium dubium.*

Throughout Northern Skye there are numerous basalt or dolerite cliffs, which have an interesting and often unexpected flora. Species of interest on such cliffs include *Orobanche alba, Vicia orobus, Saxifraga hypnoides, S. oppositifolia, Arctostaphylos uva-ursi, Asplenium adiantum-nigrum, Carlina vulgaris, Draba incana,* and numerous mosses, liverworts, and lichens.

In addition to the communities discussed above, there are several communities that occur widely throughout Skye irrespective of the underlying geology. These include the communities of the salt marshes that have developed at the heads of some of the more sheltered sea-lochs. The lower zone is characterized by abundant *Puccinellia maritima* with *Armeria maritima* and *Glaux maritima.* The upper zones are dominated by *Juncus gerardii* and *Festuca rubra,* with *Glaux maritima, Blysmus rufus, Carex extensa,* and *C. viridula* ssp. *viridula* var. *pulchella.* Communities with *Bolboschoenus maritimus* occur more locally. In recent years the grazing pressure on salt marshes has greatly increased, with the result that many of the species are more difficult to find and identify than they were in the 1970s. Shingle-beach communities occur frequently around the coast in suitable sites, with abundant *Rumex crispus, Atriplex* spp., *Galium aparine, Stellaria media,* and, in contrast to their fen occurrences in southern England, *Lycopus europaeus, Carex otrubae,* and *Scutellaria galericulata.*

Sand dunes are very rare on Skye, the most extensive being at Glen Brittle, where fore-dune communities occur with *Ammophila arenaria, Polygonum oxyspermum* ssp *raii, Elytrigia juncea, Cakile maritima,* and *Honckenya peploides.* Closed dune grassland occurs slightly inland, with *Carex arenaria, Sedum acre, Galium verum,* and *Thalictrum minus* ssp. *arenarium.*

Communities of weeds and ruderals occur at field edges, by derelict buildings, along roadsides, and in passing places throughout the island. Similarly numerous freshwater lochs occur throughout Skye. The lochs support a wide range of aquatic, swamp, and fen vegetation (Butterfield and Bell, 1996). Little is known about the vegetation of deep lochs, but in sheltered bays in many lochs communities dominated by *Nymphaea alba, Sparganium angustifolium,* and *Potamogeton natans* occur. These grade, with decreasing water depth, into stands dominated by *Phragmites australis, Schoenoplectus lacustris, Carex lasiocarpa,* or *C. rostrata.* These, in turn, grade into fen communities, usually dominated by *Carex rostrata* with a variety of other sedges, herbs, and bryophytes. On more exposed shores in many lochs *Littorella uniflora* and *Lobelia dortmanna* frequently grow together, with *Ranunculus flammula* and, more rarely, *Baldellia ranunculoides, Elatine hexandra,* and *Subularia aquatica.*

A remarkable plant of the lochans in the Sligachan area is *Eriocaulon aquaticum,* where it grows with *Carex rostrata, Eleogiton fluitans,* and *Menyanthes trifoliata* in many shallow peaty lochans. Its distribution today, with localities in Western Scotland and Ireland and eastern North America raises fascinating and, as yet, unsolved questions about its ecology and history. This and so many other problems of the flora and vegetation of Skye contribute to the botanical fascination of Skye and will, no doubt, provide challenges to many future generations of botanists on Skye. It is essential for this generation to do all that we can to ensure the conservation and survival of these botanical treasures for future generations to study and to enjoy.

THE VEGETATION OF RAASAY AND RONA

Although the flora of Raasay and Rona is comparatively well known, due to the recording by J. W. Heslop Harrison and S.J. Bungard, the vegetation of Raasay and Rona is still poorly known. Raasay is of considerable ecological interest, however, as much of the varied geology and topography of Skye is repeated on a smaller scale within Raasay (Figure 4). The extensive east-facing Jurassic limestone cliffs from Hallaig to Screapadal support vegetation similar, in many ways, to the east coast of Trotternish. The Raasay cliffs differ, however, in their abundance of *Dryas octopetala,* growing with *Pyrola rotundifolia, Epipactis atrorubens, Gymnadenia conopsea, Listera ovata, Coeloglossum viride, Carex flacca, Plantago maritima,* and *Thymus polytrichus.* Other montane plants, including *Cryptogramma crispa, Saxifraga aizoides, S. hypnoides, Thalictrum alpinum, Saussurea alpina, Arabis petraea, Silene acaulis, Draba incana, Persicaria vivipara, Salix herbacea,* and *Polystichum lonchitis* all occur below 400 m altitude. Natural areas of woodland are rare, with small stands of birch and hazel scrub occurring on steep slopes, with *Listera ovata, Hymenophyllum wilsonii,* and *H. tunbrigense,* the latter in its northernmost known world locality. There are mature deciduous (planted) woodlands around Raasay House and Forest Enterprise plantations in various parts of the island. *Sorbus rupicola* is a notable species of the Hallaig cliffs growing as it does in the Elgol and Suardal areas of Skye, on near vertical limestone or basalt cliff faces. Coastal salt-marsh, shingle, and cliff communities are well developed, and support interesting plants such as *Ligusticum scoticum, Anagallis minima,* and *Osmunda regalis.* Freshwater lochs can be of interest botanically with *Nuphar lutea, Potamogeton filiformis, P. perfoliatus, Sparganium natans, Utricularia australis,* and *Cladium mariscus.*

Much of the central part of Raasay is covered by *Scirpus cespitosus-Calluna* bog, *Agrostis-Festuca* grassland, and *Calluna* heather-moor. Notable species in these initially rather dull habitats include *Dactylorhiza lapponica, Listera cordata, Pyrola media, P. minor, Orthilia secunda, Eriophorum latifolium,* and *Pedicularis sylvatica* ssp. *hibernica.* Although the flora of Raasay shares many species in common with Skye, there are several curious absentees from Raasay, including *Saxifraga oppositifolia, S. stellaris, Rosa pimpinellifolia, Subularia aquatica, Galium boreale, G. verum, Carlina vulgaris,* and *Trollius europaeus,* all of which are locally frequent on Skye. Raasay does, however, have some species that, to date, are unknown on Skye - *Acaena inermis* (from New Zealand), *Cladium mariscus,* and the small Adder's-tongue *Ophioglossum azoricum.* These floristic differences emphasize the role of chance that must be so important in influencing the composition of island floras.

THE VEGETATION OF RUM

Rum (approximately 109 square km in area) has a long (48 km) and predominantly rocky coastline whereas the interior is almost entirely a complex of moorland and mountain with numerous streams and small lochs. The island was declared a National Nature Reserve in 1957 and its vegetation has been extensively studied as part of its conservation. Magnusson (1997) gives an account of the history, archaeology, natural history, and conservation of Rum. There is some planted woodland but almost all the island is treeless with fragments of native woodland and scrub surviving in only a few rocky places.

Geologically Rum is complex (Figure 4) and is dominated by an extinct Tertiary volcano. From a botanical viewpoint there are three main rock groups: acidic Torridonian sandstone and granophyre giving the usual base-poor soils; calcareous basalt and Triassic limestone giving local areas of calcareous soil; and gabbro, peridotite, and other ultra-mafic rocks which weather to give soils rich in magnesium. The sandstone is mostly in the lowlands and has a rather uniform and poor flora of *Molinia*-dominated grasslands and bogs, but around the coast *Ajuga pyramidalis* (curiously unknown from Skye) occurs and *Vicia orobus* grows in several places inland. The calcareous rocks are limited in areal extent and do not rise above 580 m. They support a rich cliff flora of arctic-alpine plants including *Dryas octopetala, Saxifraga nivalis, Poa alpina, P. glauca, Asplenium viride,* and *Thlaspi caerulescens* (only on Rum in the vice-county). Extensive herb-rich *Festuca-Argostis* grasslands are common on the slopes below the cliffs, and along the north coast there is a species-rich maritime *Calluna vulgaris* heath with, in one area, *Dryas octopetala* (Ratcliffe, 1977). Gullies in the sea-cliffs are the main habitat for the tall ferns *Osmunda regalis, Athyrium filix-femina, Dryopteris affinis,* and *D. oreades.* Shaded sandstone rocks on the north coast support the ferns *Dryopteris aemula* and *Hymenophyllum tunbrigense,* along with a variety of nationally rare oceanic bryophytes and lichens, several of which also grow on Sleat in southern Skye.

The basic and ultra-mafic igneous rocks form the highest range of hills, with Askival (811 m) as the highest point. They support a few rare alpine plants including *Minuartia sedoides, Tofieldia pusilla, Arenaria norvegica, Juncus biglumis,* and *J. triglumis,* and an abundance of *Arabis petraea.* The rock-crevice dwelling fern *Asplenium septentrionale* also occurs in its only known extant locality in the vice-county. The soils on these magnesium rocks are infertile and unusual. Deep weathering has resulted in friable brown soils of considerable depth and there is much surface instability with frequent redistribution and erosion by wind and water. There is thus an abundance of very open ground with only a sparse vegetation, including *Salix herbacea, Armeria maritima, Antennaria doica, Huperzia selago,* and the worm-like lichen *Thamnolia vermicularis.* On the steep slopes of Askival, Hallival, and Ruinsival, there is a large colony of Manx shearwaters which burrow deeply into the loose textured soils. As a result of manuring, there are areas of rich grassy swards (Ratcliffe, 1977).

The granophyre that forms the southern high mountains with Sgurr nan Gillean (763 m) is hard and acidic, and gives shallow soils that support more typical western Highland vegetation, including poor wind-blasted *Racomitrium* heath on the exposed summit areas. On the lower moorlands there is extensive wet heath and blanket bog on shallow peat, dominated by *Calluna vulgaris*, *Scirpus cespitosus*, and *Molinia caerulea*. Wetter bogs occur on deeper peat, with an abundance of *Molinia caerulea* whereas *Myrica gale* is very rare on Rum in contrast to its abundance on Skye. Pools within these blanket bogs support *Drosera intermedia*, *D. angelica*, *Utricularia intermedia*, *Rhynchospora fusca*, and *R.alba*.

Rum contains a range of mires on shallow peat that are influenced and flushed by running water, so-called soligenous mires. These can be sedge-dominated (e.g. *Carex panicea*, *C. echinata*, *C. hostiana*, *C. nigra*, *C. pulicaris*, *C. viridula* ssp. *oedocarpa*) with *Eleocharis multicaulis*, *E. quinqueflora*, *Ranunculus flammula*, *Leontodon autumnalis*, *Parnassia palustris*, *Selaginella selaginoides*, and *Dactylorhiza lapponica*; or rush-dominated (e.g. *Juncus effusus*, *J. articulatus*, *J. acutiflorus*, and a possible hybrid between *J. articulatus* and *J. acutiflorus*) with *Succisa pratensis*, *Prunella vulgaris*, and *Parnassia palustris*. In addition, on the ultra-mafic peridotite in the Harris area, *Schoenus nigricans* is very abundant where there is lateral water seepage. The vegetation on Rum dominated by *S. nigricans* is the closest equivalent to the characteristic *Schoenus*-rich vegetation on mires in western Ireland, especially Connemara (Ratcliffe, 1977).

The small lochs and lochans on Rum support a sparse assemblage of macrophytes typical of nutrient-poor waters (Farmer, 1984). Plants like *Nymphaea alba* are curiously rare. The pillwort, *Pilularia globulifera*, is a very rare and endangered species throughout Europe (it only occurs in Europe). It grows on the shallow margins of lochans and pools and was previously recorded from several lochans on Rum, but it disappeared, probably because cattle were removed from the island. When cattle were re-introduced in 1971, trampled areas at pool margins provided open mud on which pillwort could flourish. It was successfully re-introduced in 1997 (Magnusson, 1997). There are small areas of machair on coastal sands at Kilmory and Samhnan Insir. These areas have not been ploughed since the entire population of the island emigrated in 1828 and are rich in small herbs such as *Anthyllis vulneraria*, *Thymus polytrichus, Lotus corniculatus, Prunella vulgaris, Parnassia palustris, Daucus carota, Centaurium erythraea,* and *Gentianella campestris.*

A major ecological factor affecting the vegetation of Rum is grazing by sheep and red deer, the latter re-introduced for 'sport' in 1845 (Magnusson, 1997). When Rum was declared a National Nature Reserve in 1957, the island was severely over-grazed. As a result sheep were removed and one sixth of the deer population culled each year. The general condition of the vegetation improved, but the loss of sheep grazing resulted in a decline in the richness of the machair grassland. Highland cattle were introduced in 1970. The overall result has been that deer grazing has maintained the diversity of the grasslands on basic soils, whereas the increased culling policy has had a negligible influence on the unproductive grasslands and heaths (Pearman and Walker, 2004).

Rum's botany has, since the publication by Professor J.W. Heslop Harrison F.R.S. of a series of papers between 1939 and 1958 on the flora of Rum, been a topic of considerable interest, curiosity, and discussion. Although the authenticity of some of the records presented by Heslop Harrison has been in doubt since Raven's (1949) note suggesting that some species had been deliberately planted on Rum, the whole bizarre matter is now largely resolved thanks to Pearman and Walker's (2004) thorough analysis. They conclude that the records by Heslop Harrison for 13 arctic or arctic-alpine species such as *Carex bicolor, C. capitata, C. glacialis, Erigeron uniflorus,* and *Lychnis alpina* on Rum were most likely the result of deliberate introductions even though they had been verified by independent botanists. In contrast, the records of *Carex atrata, C. atrofusca, C. capillaris,* and *C. norvegica* were never verified by independent botanists. All these records, along with records for *Juncus capitatus. Polycarpon tetraphyllum,* and *Epilobium lactiflorum* are now disregarded. The article and book by Sabbagh (1999a, 1999b) provide fascinating accounts of this bizarre episode in the botany of Rum, as does the recently edited version of the report by John Raven on his visit to Rum in 1948 (Preston, 2004).

THE VEGETATION OF EIGG, MUCK, CANNA, SCALPAY, AND SOAY

The bedrock geology of these small islands (Figure 1) varies from island to island. Soay (9 square km) and Scalpay (20 square km) primarily consist of Torridonian sandstone (Figure 4). Canna and the adjacent island of Sanday (12.7 square km) consist almost entirely of Tertiary basalt, volcanic agglomerate, and dolerite. Eigg (31 square km) is comprised mainly of basaltic lavas, often forming long cliffs. The prominent feature of An Sgurr consists of a mass of columnar pitchstone rising to 393 m. Muck (6.5 square km) consists almost entirely of Tertiary basaltic lava, tuff, and volcanic agglomerate.

Although the main vegetation types on these islands are similar to those on Skye and Rum, there are interesting differences between the islands in the abundance of certain vegetation types and in the presence of certain species. Coastal vegetation types are naturally well represented on these islands, and the local abundance of *Scilla verna* in maritime turf near sea-cliff edges on Canna and Sanday is striking. The transition between salt marsh and *Iris pseudacorus* communities often supports an interesting mixture of species, including *Catabrosa aquatica, Lycopus europaeus, Ranunculus sceleratus* (only on Canna), *R. hederaceus,* and *Sparganium erectum.* Waterlogged areas in or just above the conspicuous *Iris*-dominated zone can support *Hypericum elodes, Apium inundatum* (only on Canna), *Anagallis tenella, Scutellaria minor,* and *Glyceria flutans* (Birks et al., 1991; Asprey, 1947).

Calluna-dominated moorland, wet heath, and blanket bog on shallow peat are widespread on the islands, but *Myrica gale* is curiously rare on some islands. Moorland on Scalpay provides the only known localities for *Lycopodium annotinum* in the vice-county, whereas *Carex paniculata* is confined to moorland flushes and fens on Muck (Dobson and Dobson, 1985). There is a large colony of *Cladium mariscus* on Soay, otherwise it is only known on Raasay and Rona in the vice-county.

None of these islands are particularly high and there is thus little mountain vegetation. *Arenaria norvegica* occurs on screes on Eigg, generally in drier habitats than it occupies on Rum. The basalt cliffs of Eigg support *Saxifraga oppositifolia. S. aizoides, Dryas octopetala, Galium boreale,* and *Draba incana. Ajuga pyramidalis* occurs locally on cliff ledges and open turf on Muck, Eigg, and Canna.

There is little native woodland on any of the islands, except for the dense stands of hazel scrub on steep, block-strewn slopes below the basalt cliffs on Eigg (Gilbert, 1984). Small areas of birch woodland, with some oak, hazel, ash, rowan, and holly occur locally on Soay (Barkley, 1953) and Canna (Birks *et al.* 1991).

Waste ground by crofts supports *Chrysanthemum segetum, Galeopsis speciosa,* and *Odontites verna.* Periodically damp open areas have a characteristic flora with *Anagallis minima, Isolepis setacea, I. cernua* (Muck only), and *Juncus bufonius.* Arable weeds appear to be declining because crofters are giving up growing patches of oats, grass for hay, and even potatoes. Sheep numbers are rising and their grazing effects are becoming increasingly marked, especially on shallow soils on steep basalt slopes.

Much remains to be discovered about the flora and vegetation of these islands and about changes in their flora and vegetation in response to changes in land-use. Like Skye, Raasay and Rona, and Rum, the botany of these islands will repay further investigation and careful recording.

CHECK LIST OF THE VASCULAR PLANTS OF SKYE, RAASAY, RONA, RUM, EIGG, MUCK, CANNA, SCALPAY, AND SOAY

Huperzia selago *Lycopodium selago* Fir Clubmoss; Garbhag an-t-Slèibhe
Among heather, on rock outcrops in moors, and on cliffs. Locally common. Circ BAMo. Skye (all squares), Ra (53, 54, 65), Ro, Ru, E, M, C, Sc, So.

Lycopodiella inundata *Lycopodium inundatum* Marsh Clubmoss; Garbhag Lèana
Boggy edge of Loch Meodal, Sleat, 1979; Loch Dubha, Sligachan, 1989 (Butterfield & Bell, 1996). Rare. Euro BoTe. Skye (42, 61).

Lycopodium clavatum Stag's-horn Clubmoss; Lus a'Mhadaidh-ruaidh
Isolated patches on moors. Very local. Circ BoTe. Skye (15, 24, 32, 33, 35, 42-47, 51, 52, 54, 60-62, 71, 72), Ra (53), Ru, E.

L. annotinum Interrupted Clubmoss; Lus a'Bhalgair
Four moorland sites on Scalpay, 1969 and 1981, confirming 1937 record. Rare. Circ BAMo. Scalpay only.

Diphasiastrum alpinum *Lycopodium alpinum* Alpine Clubmoss; Garbhag Ailpeach
In mountain grassland. Storr-Quiraing ridge; Red Hills; Cuillin; etc. Locally common. Circ ArMo. Skye (14, 24, 25, 32, 33, 36, 41-47, 51-53, 61, 62, 71, 72), Ra (53-55), Ru, E.

Selaginella selaginoides Lesser Clubmoss; Garbhag Bheag
Damp places and flushes on moors and hills. Locally common. Circ BoMo. Skye (all squares), Ra (53-55, 64, 65), Ro, Ru, E, M, C, Sc, So.

Isoetes lacustris Quillwort; Luibh nan Cleiteagan
Submerged in stony lochs. Locally plentiful. ESib BoMo. Skye (14, 24, 25, 32-35, 43-47, 50-53, 55, 60-62, 71, 72), Ra (53-55, 65), Ru, E, Sc, So.

I. echinospora *I. setacea* Spring Quillwort; Luibh Cleite an Earraich *
Submerged in more peaty lochs. Rare - or overlooked? Circ BoMo. Skye (25, 33, 44, 52, 62).

Equisetum hyemale Rough Horsetail, Dutch Rush; Biorag
Wet moorland, stream banks. Rare. Circ BoTe. Skye (26, 35, 41, 45, 55), Ru.

E. x **trachyodon = E. hyemale** x **E. variegatum** Mackay's
Horsetail
Banks of R. Hinnisdal, 1974. Rare. Skye (35, 45), Ru.

E. variegatum Variegated Horsetail; Earball an Eich Caol
Damp ground, An Garbh-choire, 1978, Bearreraig, 1984. Old
record in 36 requires confirmation. Circ BAMo. Skye (36, 41,
45, 55), Ru.

E. fluviatile Water Horsetail; Clois
In lochs and ditches. Locally common. Circ BoTe. Skye (all
squares except 23), Ra (53-55, 65), Ro, Ru, E, M, Sc, So.

E. fluviatile x **E. palustre = E.** x **dycei**
Abhainn à Ghlinne, Dalavil, 1991. Skye (60), Ru.

E. x **litorale = E. fluviatile** x **E. arvense** Shore Horsetail
Rare or overlooked? Skye (46, 47).

E. arvense Field Horsetail; Earball an Eich
Grassy areas, stream banks, waste areas, etc. In drier places
than *E. pratense*. Locally common. Circ WiBo. Skye (all
squares except 41), Ra (53, 54), Ro, Ru, E, M, C, Sc.

E. arvense x **E. palustre = E.** x **rothmaleri**
New to British Flora, 1971. Ditches at Kilmaluag. Rare. Skye
(47).

E. pratense Shady Horsetail; Earball an Eich Dubharach
Damp grassy banks, often by burns. Rare. Circ BoMo. Skye
(24, 25, 35, 44-47, 54-56), Ra (54), Ru, E.

E. sylvaticum Wood Horsetail; Cuiridin Coille
Damp places on moorland, and also arable ground. Locally
common. Circ BoTe. Skye (all squares except 34, 71), Ra (53,
54, 64, 65), Ro, Ru, E, M, Sc.

E. palustre Marsh Horsetail; Cuiridin
Wet places. Common. Circ BoTe. Skye (all squares except 23,
41, 53), Ra (53-55, 64, 65), Ro, Ru, E, M, Sc, So.

E. palustre x **E. telmateia = E.** x **font-queri**
Damp ground and ditches, both sides of Staffin road, along a
two mile stretch. Locally plentiful. Skye (55, 56).

E. telmateia Great Horsetail; Earball an Eich Mòr
Most records from Trotternish, on Jurassic rocks. Locally
common. Euro SoTe. Skye (24-26, 36, 44-47, 54-56), Ra (53,
54), Ru.

Ophioglossum vulgatum Adder's-tongue; Teanga na Nathrach
In limestone grikes in Allt nan Leac valley and other sites, (51); N. of Suardal (62). Rare. Should be looked for in other basic areas. Circ Temp. Skye (44, 51, 61, 62, 71), Ra (53, 54, 64), Ru, E.

O. azoricum Small Adder's-tongue; Teanga na Nathrach Beag
In short maritime turf, Eilean Tigh, and on Fladday, Raasay, 1996, and S. Rona, 1999. SubO BoTe. Ra (54, 55, 65), Ro, Ru.

Botrychium lunaria Moonwort; Lus nam Mìos
Easily overlooked in dry turf (including roadside verges), and scree on hills. Still not recorded in much of Sleat. Circ BoTe. Skye (14, 15, 24, 25, 32, 33, 36, 37, 42-47, 50-55, 62), Ra (53-55), Ru, E, M.

Osmunda regalis Royal Fern; Raineach Rìoghail
Stony and peaty edges of lochs; rocky stream banks; sea-cliff ledges. Local, commoner in S. Skye. SubO SoTe. Skye (15, 26, 33, 41, 42, 47, 50-52, 60-62, 72), Ra (53-55), Ru, Sc, So. (Planted in Skye 24, 35, 43).

Cryptogramma crispa Parsley Fern; Raineach Pheirsill
Scattered plants in scree and among boulders on hills. Locally plentiful, Red Hills; very rare, Storr. Euro BoMo. Skye (41, 42, 45, 52-54, 62, 71, 72), Ra (53, 54), Ru.

Pilularia globulifera Pillwort; Feur A'Phiobair
Formerly at Harris, Rum. Re-introduced there 1990s. Rare. SubO Temp. Ru.

Hymenophyllum tunbrigense Tunbridge Filmy-fern; Raineach Còinnich Fiaclach
Damp, shady rocks, usually in lowland woodland and wooded ravines. Tokavaig woods; Kyleakin; Raasay; etc. Ocea Temp. Skye (50, 53, 60, 61, 71, 72), Ra (53, 54), Ru, Sc.

H. wilsonii Wilson's Filmy-fern; Raineach Còinnich
Damp rocks, mossy boulders, tree trunks—in less shade and extends to higher altitudes than above. Locally plentiful. Ocea BoTe. Skye (all squares), Ra (53-55), Ro, Ru, E, C, Sc, So.

Trichomanes speciosum Killarney Fern; Raineach Chillearnaidh
Filamentous gametophyte stage only, in deeply shaded caves and crevices in sandstone walls of ravines and coastal cliffs. Rare, probably overlooked. Ocea Temp. Skye (61, 71, 72).

Polypodium vulgare agg. Polypodies; Clach-raineach Chaol
On rocks and trees, locally common. Circ BoTe. Skye (all squares), Ra (53-55, 64, 65), Ro, Ru, E, M, C, Sc, So. Records need separating into

P. vulgare s.s. (Polypody; Euro BOTe) Skye (25, 33, 41, 44, 45, 51, 55, 60-62), Ru, E; and **P. interjectum** (Intermediate Polypody; SubO Temp) Skye (51, 52), Ra (53, 54), M, Sc.

Pteridium aquilinum Bracken; Raineach
Heaths, woodland and former grassland. Common. Circ Temp. Skye (all squares), Ra (53-55, 64, 65), Ro, Ru, E, M, C, Sc, So.

Phegopteris connectilis *Thelypteris phegopteris* Beech Fern; Raineach Fhaidhbhile
Damp rock ledges and woods. Local. Circ BoTe. Skye (all squares except 37 and 56), Ra (53-55, 64, 65), Ro, Ru, E, Sc, So.

Oreopteris limbosperma *Thelypteris oreopteris* Lemon-scented Fern, Mountain Fern; Crim-raineach
Rough grassland, scree, etc., not confined to mountains. Common. Euro Temp. Skye (all squares), Ra (53, 54, 64, 65), Ro, Ru, E, C, Sc, So.

Phyllitis scolopendrium *Asplenium scolopendrium* Hart's-tongue; Teanga an Fhèidh
Frequent in areas of Jurassic rocks or Cambrian limestone. Euro Temp. Skye (all squares except 23, 25, 33, 37, 42, 43, 46), Ra (53, 54, 65), Ru, E, C, Sc.

Asplenium adiantum-nigrum Black Spleenwort; Raineach Uaine
Rocks and walls. Common. Euro Temp. Skye (all squares), Ra (53-55, 64, 65), Ro, Ru, E, M, C, Sc, So.

A. marinum Sea Spleenwort; Raineach na Mara
Rock crevices and sea cliffs. Missing from areas where rivers make sea-lochs less saline, e.g. upper arms of Loch Snizort. Local. SubO SoTe. Skye (all squares except 43-45), Ra (53-55, 64, 65), Ro, Ru, E, M, C, Sc, So.

A. trichomanes agg. Maidenhair Spleenwort; Dubh-chasach
On rocks and walls. Circ SoTe. Skye (all squares), Ra (53-55, 64, 65), Ro, Ru, E, M, C, Sc, So. Check for ssp. **quadrivalens** (42, 45 only so far) as most records in agg. may be this.

A. viride *A. trichomanes-ramosum* Green Spleenwort; Ur-thalamhainn
Crevices of limestone or ultra-basic rocks. Locally plentiful in Suardal area; rare on Storr-Quiraing ridge. Circ BoMo. Skye (26, 32, 33, 43-47, 51, 52, 55, 60-62, 72), Ra (53, 54), Ru, Sc.

A. ruta-muraria Wall-rue; Rù Bhallaidh
Basic rocks and old walls. Occasional; frequent in limestone areas. Circ Temp. Skye (14, 15, 24, 25, 32, 33, 35, 41, 42, 44-46, 50-52, 54, 56, 60-62, 72), Ra (53, 54, 65), Ro, Ru, E, M, C, Sc, So.

A. septentrionale Forked Spleenwort; Lus a'Chorrain Gòbhlach *
Rock crevices. Old record from Hallaig, Raasay, not confirmed. Papadil, Rum. Euro Temp. Ru.

Ceterach officinarium Rustyback; Raineach Ruadh
In 'ballast' of old railway track, Suardal. Introduced. SubM SubA. Skye (62), old wall, Eigg. E.

Athyrium filix-femina Lady-fern; Raineach Moire
Rough ground, woods, and among rocks on hills. Common. Circ BoTe. Skye (all squares), Ra (53-55, 64, 65), Ro, Ru, E, M, C, Sc, So.

Gymnocarpium dryopteris *Thelypteris dryopteris* Oak Fern; Sgeamh Dharaich
Rocks, scree, damp woods. Rare. Circ BoTe. Skye (23, 24, 26, 34, 42, 45, 46, 50-53, 60, 61, 71, 72), Ra (53, 54), Ru, E, Sc.

Cystopteris fragilis Brittle Bladder-fern; Frith-raineach
Rock crevices and walls. Widespread, usually on basic rock. Circ WiBo. Skye (all squares except 25, 37, 50, 56), Ra (53-55, 64, 65), Ru, E.

Woodsia alpina Alpine Woodsia; Raineach Mhion Ailpeach
Basic cliffs. Old record (1884) Quirang. Cliffs above Loch Cuithir, 2001. Rare. Circ BAMo. Skye (45, 46).

Polystichum setiferum x **P. aculeatum** = **P. x bicknellii**
Foot of old wall, Torrin, 1976. Skye (52, 62).

P. aculeatum Hard Shield-fern; Ibhig Chruaidh
Rock crevices and boulder scree. Frequent, but more plentiful on limestone or ultra-basic rock. EAsi Temp. Skye (all squares except 15, 34, 37), Ra (53-55, 64, 65) Ru, E, C, Sc, So.

P. lonchitis Holly-fern; Raineach Chuilinn
Basic rock crevices and boulder scree. Blaven (comparatively frequent); Storr (rare); Quiraing; Suardal; Raasay. Circ BoMo. Skye (45-47, 51-52, 61, 62), Ra (53, 54).

Dryopteris oreades *D. abbreviata* Mountain Male-fern; Marc-raineach Ailpeach
Mountain scree - Cuillin; Blaven; Quiraing. Very local. SubO BoMo. Skye (42, 46, 51, 52, 62, 71, 72), Ra (54), Ru.

D. filix-mas agg. Male-fern; Marc-raineach
   Damp hillsides, rocks and woods. Common. Circ Temp. Skye
   (all squares), Ra (53-55, 65), Ro, Ru, E, M, C, Sc, So.
D. affinis Scaly Male-fern; Mearlag
   Hillsides, grassland, and woodland edges. Common. Euro
   Temp.
   Skye (all squares), Ra (53, 54, 64, 65), Ro, Ru, E, M, C, Sc, So,
   but records of *D. filix-mas* s.s. still need revision.
D. aemula Hay-scented Buckler-fern; Raineach Phreasach
   Woods and shady rocks. Local. Ocea Temp. Skye (14, 24, 26,
   32, 33, 37, 41, 43, 46, 47, 50-53, 60-62, 71, 72), Ra (53-55,
   64, 65), Ro, Ru, E, Sc, So.
D. carthusiana *D. spinulosa* Narrow Buckler-fern; Raineach Caol
   Wet woods and heaths. Rare. ESib BoTe. Skye (25, 37, 44, 61,
   72), Ra (53, 54), Ru, E, M, So.
D. dilatata *D. austriaca* Broad Buckler-fern; Raineach nan Radan
   Shady rock ledges, boulder scree, woods. Local. Euro Temp.
   Skye (all squares except 56), Ra (53-55, 64, 65), Ro, Ru, E, M,
   C, Sc, So.
D. expansa *D. assimilis* Northern Buckler-fern; Raineach
   Thuathach
   Wall of Dun Fiadhairt, near Dunvegan, 1971. Elsewhere? Circ
   BoMo. Skye (25, 71), Ra (53), E, M.
Blechnum spicant Hard-fern; Raineach Chruaidh
   Moorland and rock ledges on hills. Common. Euro Temp. Skye
   (all squares), Ra (53-55, 64, 65), Ro, Ru, E, M, C, Sc, So.
Juniperus communis ssp. **communis** Juniper; Aitinn
   Erect, but growing downwards, on a sea-cliff at Rigg; also on
   Beinn Bhuidhe. Rare, Skye only. ESib BoTe. Skye (55, 72).
J. communis ssp. **nana** *J. communis* ssp. *alpina* Dwarf Juniper;
   Iubhar Beinne
   Procumbent on stony ground, or cliff ledges on hills; also on
   sea cliffs. Local. Circ BoMo. Skye (all squares except 34, 36,
   37, 44, 45, 54), Ra (53-55, 64, 65), Ro, Ru, E, M, C, Sc, So.
Nymphaea alba ssp. **alba** White Water-lily; Duilleag-bhaite Bhàn
   Not in all lochs. Locally common. Euro Temp. Skye (23, 25, 26,
   33, 34, 41-44, 47, 50-53, 60-62, 71, 72), Ra (53-55, 64), Ro,
   Ru, C (introduced), Sc, So.
N. alba ssp. **occidentalis** Lesser White Water-lily
   Three lochs in S Skye, and may occur in others. Intermediates
   with ssp. *alba* can occur. Euro Temp. Skye (41, 50, 62). Ra,
   (unlocalised) So.

Nuphar lutea Yellow Water-lily; Duilleag-bhaite Bhuidhe
Loch A'Mhuilinn, Scalpay. Introduced? Recently refound in Loch na Leanna (Ra 54) and Loch Mor on Fladday. ESib BoTe. Ra (54, 55), Sc.

Caltha palustris Marsh-marigold; Lus Buidhe Bealltainn
Common in marshy ground, with ssp. **radicans** (spp. *minor*) on mountains. Circ WiBo. Skye (all squares), Ra (53-55), Ro, Ru, E, M, C, Sc, So.

Trollius europaeus Globeflower; Leolaicheann
In damp grassland, and on rock ledges on hills. Locally plentiful. Euro BoMo. Skye (all squares except 23, 37), Ru, E. Missing from Raasay.

Anemone nemorosa Wood Anemone; Flùr na Gaoithe
Woodland, but more often in open grassland. Local. ESib Temp. Skye (all squares except 15, 23, 26, 37), Ra (53-55, 64), Ru, E, Sc, So.

Ranunculus acris Meadow Buttercup; Buidheag an-t-Samhraidh
Common in grassland. EAsi WiBo. Skye (all squares), Ra (53-55, 64, 65), Ro, Ru, E, M, C, Sc, So.

R. repens Creeping Buttercup; Dithean Buidhe
A vigorous arable weed. Common. EAsi BoTe. Skye (all squares except 41), Ra (53-55, 64, 65), Ro, Ru, E, M, C, Sc, So.

R. bulbosus Bulbous Buttercup; Fuile-thalamhainn
Thought to be introduced (Dunvegan, Skeabost), until found in dune grassland, Glen Brittle. Rare. Euro SoTe. Skye (24, 34, 42, 44), Ru, M, C.

R. auricomus Goldilocks Buttercup; Gruag Moire
In hazel scrub, Staffin, 1976. Rare. Euro BoTe. Skye (46).

R. sceleratus Celery-leaved Buttercup; Torachas Biadhain
Marshy ground above the shore, Sanday (Canna) only. Circ BoTe. C.

R. flammula Lesser Spearwort; Glaisleun
Common. Euro Temp. Skye (all squares), Ra (53-55, 64, 65), Ro, Ru, E, M, C, Sc, So. Critical work is needed to see whether spp. **minimus** occurs in exposed places near the sea; ssp. **scoticus** is recorded from loch shores in Skye (42, 43, 53), Ra (53, 54), E.

R. ficaria Lesser Celandine; Brog na Laireadh
Common. So far most plants examined have been ssp. **ficaria**; ssp. **bulbifer** only in 'big house' gardens (24, 35, 44, 45). Euro SoTe. Skye (all squares), Ra (53-55, 64, 65), Ro, Ru, E, M, C, Sc, So.

R. hederaceus Ivy-leaved Crowfoot; Fleann Uisge Eidheannach
On wet muddy tracks or in shallow water, Talisker area. Rare. SubO SoTe. Skye (31-33, 50), Ru, E, M, C.

R. trichophyllus Thread-leaved Water-crowfoot; Lion na h-Aibhne
Edges of stream joining Loch Fada and Loch Leathan. Plentiful or completely absent according to water level. Circ WiBo. Skye (45).

Thalictrum minus ssp. **arenarium** Lesser Meadow-rue; Rù Beag
Dune grassland at Glen Brittle. Rare due to lack of suitable habitat. EAsi BoTe. 42 only on Skye, but Ru, E, M, C.

T. alpinum Alpine Meadow-rue; Rù Ailpeach
Damp rock ledges and scree; damp grass, edges of burns. Locally common in hilly areas. Circ ArMo. Skye (14, 15, 23-25, 32, 33, 41-47, 51-55, 61-62, 71, 72), Ra (53, 54), Ru, E, Sc.

Papaver dubium Long-headed Poppy; Currag an Righ
Introduced? Casual only? ESib SoTe. Recorded in the past from Skye (47, 52, 61). More recently from E, M.

Meconopsis cambrica Welsh Poppy; Crom-lus Cuimreach
Garden escape in various places, including Portree and Dunvegan. Introduced. Ocea BoTe. Skye (15, 24, 35, 44, 51, 52, 60, 61, 72), Ra (53), M.

Pseudofumaria lutea *Corydalis lutea* Yellow Corydalis; Fliodh an Tughaidh Buidhe
Established on wall of Episcopal Church, Portree. Introduced. Skye (44).

Ceratocapnos claviculata *Corydalis claviculata* Climbing Corydalis; Fliodh an Tughaidh
Woods and among boulders in shade, Sleat and Torrin. Rare. Ocea Temp. Skye (52, 60, 71, 72), Ra (53, 54, 64), Sc, So.

Fumaria capreolata White Ramping-fumitory; Fuaim an t-Siorraimh Bàn
Garden weed for many years in Kyleakin. Appeared in Aird, Sleat, 1978. Introduced. SubM SubA. Skye (50, 72), Ru, M.

F. bastardii Tall Ramping-fumitory; Fuaim an t-Siorraimh Ard
Arable weed on Eigg up to 1970s. Med Atl.

F. muralis Common Ramping-fumitory; Fuaim an t-Siorraimh Coitcheann

Croft weed, Uigshader. Introduced. Ocea SoTe. Skye (44). M.

F. officinalis Common Fumitory; Lus Deathach-thalamhainn

Arable weed, formerly at Torrin; also Tormore, Sleat; Portree (1997). Rare. Introduced. Euro SoTe. Skye (44, 52, 60), E.

Ulmus glabra Wych Elm; Leamhan

Often only a single specimen, in woodland or on cliff ledges. Widespread but rare. Euro Temp. Skye (all squares except 23, 26, 34, 37, 56), Ra (53, 54), Ro, Ru, E, M, C.

Humulus lupulus Hop; Lus an Leanna

Dunvegan Castle grounds; Viewfield, Portree; also 25. Rare. Introduced. ESib Temp. Skye (24, 25, 44), Ru.

Urtica dioica Common Nettle; Deanntag

Waste ground, woods, and round houses and out-buildings. Often marks the site of old crofts. Common. ESib BoTe. Skye (all squares), Ra (53-55, 64, 65), Ro, Ru, E, M, C, Sc, So.

U. urens Small Nettle; Deanntag Bhliadhnail

Waste places and cultivated ground. Rare - or overlooked? ESib SoTe. Skye (44, 56, 60, 61, 71), Ru, E, M.

Parietaria judaica Pellitory-of-the-wall; Lus a'Bhalla

Wall of Gun Court, Dunvegan Castle. Introduced. SubM SubA. Skye (24).

Myrica gale Bog-myrtle; Roid

Wet moorland. Locally common. SubO BoTe. Skye (all squares except 23, 37), Ra (53-55, 64, 65), Ro, Ru, E, M, C (introduced), Sc, So.

Quercus petraea Sessile Oak; Darach

Commoner than *Q. robur*, especially in Sleat. Both missing from Trotternish and NW Skye. Local. Euro Temp. Skye (41, 42, 50, 51, 53, 60, 61, 71, 72), Ra (53), Ru, E, M (introduced), C, Sc, So.

Q. robur Pedunculate Oak; Darach Gasagach

Recorded mainly in S Skye, and in Raasay and Scalpay. Local. Euro Temp. Skye (24, 35, 50, 51, 60, 61, 71, 72), Ra (53-55), Ro, E, Sc, So.

Betula pendula Silver Birch; Beith Dhubhach

Less common than *B. pubescens.* ESib BoTe. Skye (15, 24, 25, 35, 36, 41, 43-45, 47, 50-53, 55, 56, 60-62, 71, 72), Ra (53, 54, 65), Ru, E, M (introduced), C, So.

B. pubescens Downy Birch; Beith Charraigeach
Locally common. Numbers and growth reduced by sheep and muirburn. Check whether ssp. **pubescens** or ssp. **tortuosa** occurs. ESib BoTe. Skye (all squares except 14, 37), Ra (53-55, 64, 65), Ro, Ru, E, Sc, So.

Alnus glutinosa Alder; Fèarna
Wet places, often along streams. Very local. ESib Temp. Skye (14, 24, 25, 33-36, 44-47, 50-53, 55, 60-62, 71, 72), Ra (53, 54), Ro, Ru, E, M (introduced), C, Sc.

Corylus avellana Hazel; Caltainn
Locally plentiful, forming belts of woodland below bands of basalt cliffs and Jurassic limestone. Size and numbers reduced by sheep. Euro Temp. Skye (all squares), Ra (53-55, 64, 65), Ro, Ru, E, M, C, Sc, So.

Chenopodium album Fat-hen; Càl Slapach
Waste places and shingle. Perhaps under-recorded, or commoner in the past. EAsi WiTe. Skye (35, 41-46, 51, 52, 61, 62, 72), Ra (53), Ru, E, C.

Atriplex prostrata, *A. hastata* Spear-leaved Orache; Praiseach Mhin Leathann
Seashores on sand, shingle, and mud. Skye specimens do not match those from further south. More work is required. Locally plentiful. ESib WiTe. Skye (24, 33-36, 43-45, 47, 50-52, 62, 71), Ra (53-55, 64, 65), Ro, Ru, E, M, C, Sc.

A. glabriuscula Babington's Orache; Praiseach Mhin Chladaich
Easily confused with above, perhaps less common. SubO BoTe. Skye (14, 15, 24-26, 32-36, 44, 45, 50-52, 60-62), Ra (53, 65), Ro, Ru, E, M, C, So.

A. praecox Early Orache; Praiseach Mhin Thràth
Small reddish plants among shingle, low down on the beach. New to British Flora, 1975 (W. Ross and W. Sutherland). Skye 1978. Local. Euro BoMo. Skye (25, 32, 34, 36, 45, 52, 53, 61, 71), Ra (65), Ro, E.

A. patula Common Orache; Praiseach Mhin Chaol
The least common! Essential to sort out. *A. patula* is (? was) a weed of cultivated ground. ESib WiTe. Skye (35-37, 41, 44, 47, 61, 62, 72), Ra (53), E, M, C, Sc, So.

A. laciniata Frosted Orache; Praiseach Mhin Airgeadach
Recorded from Loch Brittle in 1902 and still there in 1992. Ocea Temp. Skye (42), E.

Beta vulgaris ssp. **maritima** Sea Beet; Biatas Mara
2-3 plants on Glen Brittle beach, 1972. Casual? Skye (42), Muck, 2000.

Salicornia europaea Common Glasswort; Lus na Glainne
In most salt-marshes. Local. Circ WiTe. Skye (24, 33, 45, 51-53, 60-62, 71, 72), C.

Suaeda maritima Annual Sea-blite; Praiseach na Mara
In salt-marshes and on seashores, less common where sea is 'freshened' by river water. EAsi SoTe. Skye (24, 33-35, 44, 45, 51-53, 60-62, 71, 72), Ra (53), C.

Salsola kali ssp. **kali** Prickly Saltwort; Lus an t-Salainn
Camas Ban, Portree - not seen recently. Rare due to lack of sandy shores on Skye. ESib SoTe. Skye (44), Ru, E.

Claytonia sibirica Pink Purslane; Seachranaiche
In estate woodland - Dunvegan, Portree, Tote, Kilmarie, Raasay, and elsewhere. Introduced. Skye (15, 24, 25, 44, 46, 51, 60, 61, 71), Ra (53).

Montia fontana Blinks; Fliodh Uisge
Common in wet places; can also be a garden weed. Ssp. **fontana** and ssp. **variabilis** have been recorded. Further work required. Euro BoTe. Skye (all squares), Ra (53-55, 64, 65), Ro, Ru, E, M, C, Sc, So.

Arenaria serpyllifolia Thyme-leaved Sandwort; Lus nan Naoi Alt Tiomach
Ord, 61; garden weed, Kyleakin, 72. Rare. ESib SoTe. Skye (24, 33, 61, 72), Ru, E, M, C.

A. norvegica ssp. **norvegica** Arctic Sandwort; Lus nan Naoi Alt Lochlannach
Bare gravelly ground on Ruinsval, Rum; and on Eigg. Rare. Euro ArMo. Ru, E.

A. balearica Mossy Sandwort
Naturalised near Raasay House; also at Kilmarie, Strathaird; and perhaps other 'big house' gardens. Introduced. Skye (51), Ra (53).

Moehringia trinervia Three-nerved Sandwort; Lus nan Naoi Alt Trì-lèitheach
Woodland, Balachuirn, Raasay, 1979 and along E. coast, 1994. Confirms 1930s records for Raasay. Rare. Euro Temp. Ra (53, 54, 64, 65), Ru, E.

Honckenya peploides Sea Sandwort; Lus a'Choill
In sand and sandy shingle. Local. Circ WiBo. Skye (14, 24-26, 32-34, 36, 37, 42, 44-46, 50, 51, 53, 60-62, 72), Ru, E, M, C, Sc.

Minuartia sedoides *Cherleria sedoides* Cyphel; Lus an Tuim Chòinnich
Cushions in rocky ground on Storr-Quiraing ridge; gravel at sea level, mouth of R. Haultin, 45. Not yet found on the Cuillin. Euro ArMo. Skye (45, 46, 55), Ru.

Stellaria media Common Chickweed; Fliodh
Weed of cultivated ground, dominant in wet summers. EAsi WiTe. Skye (all squares except 23, 41), Ra (53-55, 64, 65), Ro, Ru, E, M, C, Sc, So.

S. holostea Greater Stitchwort; Tursach
Grassy banks and scrub woodland. Very local. ESib Temp. Skye (24, 25, 33, 34, 36, 41, 44-46, 50, 51, 53, 54, 60-62, 71, 72), Ra (53, 54), Ro, Ru, E, M, Sc.

S. graminea Lesser Stitchwort; Tursarain
Isolated patches in grassland. Rare. ESib BoTe. Skye (24, 26, 33-36, 45, 46, 50, 52, 61, 71, 72), Ra (53), Ru, E, M, Sc.

S. uliginosa *S. alsine* Bog Stitchwort; Flige
Beside burns and springs, and in ditches. Locally common. Euro Temp. Skye (all squares except 41), Ra (53-55, 64, 65), Ro, Ru, E, M, C, Sc, So.

Cerastium arcticum Arctic Mouse-ear; Cluas Luch Artach
Wet rock ledges and scree, Cuillin hills; wet stony ground, Storr; Blaven and Garbh-bheinn, 1999. Rare. Euro ArMo. Skye (42, 45, 52).

C. fontanum *C. holosteoides* Common Mouse-ear; Cluas Luch
Common. ESib BoTe. Skye (all squares), Ra (53-55, 64, 65), Ro, Ru, E, M, C, Sc, So.

C. glomeratum Sticky Mouse-ear; Cluas Luch Fhàireagach
Common arable weed. Euro SoTe. Skye (all squares except 41-43, 46, 53), Ra (53, 54, 65), Ro, Ru, E, M, C, Sc, So.

C. diffusum *C. atrovirens* Sea Mouse-ear; Cluas Luch Mara
Records mainly from west side of Skye - overlooked elsewhere? Euro Temp. Skye (14, 23-26, 32, 33, 35-37, 42, 46, 47, 60), Ra (55), Ro, Ru, E, M, C, So.

C. semidecandrum Little Mouse-ear; Cluas Luch Bheag *
Dry, sandy soils. Rare, or overlooked? Euro Temp. Skye (50, 60), Ru, C.

Sagina nodosa Knotted Pearlwort; Mungan Snaimte
Wet rocks, gravel, or grass, near the sea. Easily overlooked when not in flower. ESib BoTe. Skye (33, 36, 42, 51, 61, 62), Ru, E, M, C.

S. subulata Heath Pearlwort; Mungan Mòintich
Dry gravelly places on moorland and hills. Local. Euro Temp. Skye (14, 15, 23-25, 31-36, 41, 42, 44-47, 51-56, 60-62, 71), Ra (53-55, 64), Ru, E, M, C, Sc, So.

S. saginoides Alpine Pearlwort; Mungan Ailpeach
Rare in fine scree, Storr and Quiraing areas. Circ ArMo. Skye (45-47).

S. procumbens Procumbent Pearlwort; Mungan Làir
Paths, bare ground, grassy banks. Garden weed. Common. ESib BoTe. Skye (all squares), Ra (53-55, 64, 65), Ro, Ru, E, M, C, Sc, So.

S. apetala Annual Pearlwort; Mungan *
Not common. Possible confusion with *S. procumbens*. Euro SoTe. Skye (33, 35, 47, 51, 61), Ru.

S. maritima Sea Pearlwort; Mungan Mara
A seashore plant, probably under-recorded. Euro SoTe. Skye (15, 24, 33, 34, 44, 51, 52), Ra (53). Canna 2001.

Spergula arvensis Corn Spurrey; Corran Lìn
A common and infuriating weed of cultivated ground. ESib WiTe. Skye (all squares except 23, 37, 41, 53, 54, 56), Ra (53, 54), Ru, E, M, C, Sc, So.

Spergularia rupicola Rock Sea-spurrey; Corran na Creige
Coastal rocks - recorded once from Rum. Ocea Temp. Ru.

S. media Greater Sea-spurrey: Corran Mara Mòr
Salt-marshes. Local. Needs separating from *S. marina*. ESib SoTe. Skye (24, 33-35, 44, 45, 50-53, 60-62, 71, 72), Ro, Ru, E, C, Sc.

S. marina Lesser Sea-spurrey; Corran Mara Beag
Less common - or overlooked? Check petals and seeds. Circ SoTe. Skye (14, 24, 25, 37, 43, 45, 50, 52, 60-62, 71, 72), M, C, Sc, So.

S. rubra Sand Spurrey; Corran Gainmhich
Gravel by bridge, Kinloch, 71; shore at Eynort, 32. Introduced? Rare. Euro SoTe. Skye (32, 36, 47, 62, 71), Ra (54).

Lychnis flos-cuculi Ragged-Robin; Caorag Lèana
Wet places, locally common. ESib Temp. Skye (all squares), Ra (53-55, 65), Ro, Ru, E, M, C, Sc, So.

Silene uniflora *S. martima* Sea Campion; Coirean na Mara
Sea cliffs, shingle, and mountain ledges. Local. SubO BoTe. Skye (all squares except 60, 71), Ra (53, 54, 64), Ru, E, M, C, So.

S. acaulis Moss Campion; Coirean Còinnich
Mountains and sea cliffs. Descends to sea level at Rubha Hunish. Locally plentiful. Euro ArMo. Skye (14, 15, 23-26, 33, 36, 41-47, 51-55), Ra (53, 54), Ru, E, C.

S. latifolia *S. alba* White Campion; Coirean Bàn
Croft casual, Prabost, 1970-72; perhaps only casual elsewhere. ESib SoTe. Skye (36, 44, 45, 60), Ru, M.

S. dioica Red Campion; Cirean Coilich
Sea cliffs, mountain ledges, river banks, woods. Very local. Euro BoTe. Skye (all squares), Ra (53, 54), Ru, E, M, C, Sc, So.

Persicaria bistorta *Polygonum bistorta* Common Bistort; Biolur
Dunvegan Castle grounds, 24; also 42. Introduced. Skye only. EAsi BoTe. Skye (24, 42).

P. vivipara *Polygonum viviparum* Alpine Bistort; Biolur Ailpeach
Isolated patches in damp grassland at a range of altitudes from 150ft; dwarf specimens in summit grassland, Storr; Cuillin corries. Local. Circ BAMo. Skye (14, 15, 24, 25, 32, 36, 42, 44-47, 51-53, 55, 61, 62, 71, 72), Ra (53), Ru.

P. amphibia *Polygonum amphibium* Amphibious Bistort; Gluineach an Uisge
All records are the terrestrial form, from damp ground, not even near water except in 33. Very local; commoner in S. Skye. Circ BoTe. Skye (24-26, 33, 42, 50-52, 60-62), E, M, C, So.

P. maculosa *Polygonum persicarium* Redshank, Persicary; Gluineach Dhearg
Cultivated ground and waste places. Common. EAsi Temp. Skye (all squares except 23, 41, 54-56), Ra (53, 54), Ro, Ru, E, M, C, Sc, So.

P. hydropiper *Polygonum hydropiper* Water-pepper; Gluineach Theth
In wet ground, or in ditches. Can be identified by chewing a leaf - tastes peppery! Occasional. Circ Temp. Skye (14, 15, 24, 25, 32, 33, 35, 36, 42, 44-47, 50-53, 60-62, 71, 72), Ra (53-55), Ro, Ru, E, M, C, Sc, So.

Koenigia islandica Iceland-purslane; Cainigidhe

In damp gravel or fine scree on Trotternish hills from Ben Dearg to Ben Edra. Altitudes from 800ft, S of Loch Cuithir, to 2350ft near summit of Storr. So far not found in Quiraing area, or on Cuillin hills. Locally plentiful annual. Circ ArMo. Skye (44-46, 55).

Polygonum oxyspermum ssp. **raii** Ray's Knotgrass; Gluineach na Tràighe

Sandy shores. Glen Brittle; Rubha Sloc an Eorna, Sleat; Rum; Eigg. Rare. Euro WiTe. Skye (33, 42, 50), Ru, E.

P. arenastrum Equal-leaved Knotgrass; Gluineach Ghainmhich *

Weed on paths, Muck; Eyre, Raasay. Probably elsewhere. EAsi WiTe. Ra (53), M.

P. aviculare Knotgrass; Gluineach Bheag

A fairly common weed of waste and arable ground. Circ WiTe. Skye (all squares except 23, 41, 54), Ra (53, 54), Ru, E, M, C, Sc, So.

Fallopia japonica *Polygonum cuspidatum* Japanese Knotweed; Gluineach Sheapanach

Cultivated in estate woodland and often escapes. Introduced. Skye (14, 24, 32, 33, 35, 44, 45, 51, 52, 60-62, 71, 72), Ra (53), Ru, So.

F. convolvulus *Polygonum convolvulus* Black-bindweed; Gluineach Dhubh

In shingle where Gillean Burn enters the sea (Sleat) in 1960s; Glen Brittle 1972; arable weed, Uigshader and Lynedale. Rare. ESib WiTe. Skye (35, 42, 44, 50), Ra (53), Ru, E, M, Sc.

Rumex acetosella agg. Sheep's Sorrel; Sealbhag nan Caorach

Rough grassland, and a weed of cultivated ground. Common. ESib WiTe. Skye (all squares), Ra (53-55, 64, 65), Ro, Ru, E, M, C, Sc, So.

R. acetosa Common Sorrel; Sealbhag

Grassland, hills, and woods. Common. ESib BoTe. Skye (all squares), Ra (53-55, 64, 65), Ro, Ru, E, M, C, Sc, So.

R. longifolius Northern Dock; Copag Thuathach

Damp places, ditches. Probably under-recorded. EAsi BoMo. Skye (14, 15, 24, 26, 33, 35, 44, 45, 51, 52, 62, 71, 72), Ra (53, 54), Ru, C.

R. crispus Curled Dock; Copag Chamagach

Often in shingle and on rocky shores. ESib SoTe. Skye (all squares), Ra (53-55, 64, 65), Ro, Ru, E, M, C, Sc, So.

R. conglomeratus Clustered Dock; Copag Bhagaideach
Damp grassy places. Records mainly in NW Skye. ESib SoTe.
Skye (14, 23-25, 33-35, 55, 56, 60, 61). E?

R. sanguineus Wood Dock; Copag Coille
Woodland, Armadale; edge of wood, shore at N. Fearns,
Raasay. Rare. Euro Temp. Skye (51, 60-62), Ra (53, 54).

R. obtusifolius Broad-leaved Dock; Copag Leathann
Waste and cultivated ground. Should be in all squares.
Common. Euro Temp. Skye (all squares except 41, 42, 54), Ra
(53-55, 64, 65), Ro, Ru, E, M, C, Sc, So.

Oxyria digyna Mountain Sorrel; Sealbhag nam Fiadh
Wet mountain rock ledges - Trotternish, Cuillin, Blaven, etc.
On gravel at sea level at Torrin. Local. Circ ArMo. Skye (24, 26,
32, 33, 41-47, 51, 52, 54, 55, 71, 72), Ra (54), Ru.

Armeria maritima Thrift; Neòinean Cladaich
Salt-marshes and sea cliffs; also on mountains. Locally
common. Circ WiBo. Skye (all squares), Ra (53-55, 64, 65), Ro,
Ru, E, M, C, Sc, So.

Elatine hexandra Six-stamened Waterwort; Bosan na Dìge
Rare on fine gravel under water in lochs in Talisker area, 1973.
Other lochs in Skye added since. Loch Papadil, Rum, 1947.
Easily overlooked. Euro Temp. Skye (32-34, 46, 61, 72), Ru.

Hypericum androsaemum Tutsan; Meas an Tuirc Coille
Scattered plants on rock and cliff ledges, and in woods. SubM
SubA. Skye (24, 26, 32, 33, 41-44, 50-54, 60-62, 71, 72), Ra
(53, 54), Ru, E, M, C, Sc, So.

H. tetrapterum Square-stalked St John's-wort; Beachnuadh
Fireann
Most records from marshy ground near the sea, and on
Jurassic rocks or gneiss. Euro Temp. Skye (14, 24-26, 32, 33,
35-37, 44, 47, 51, 55, 56, 60, 61), Ra (54), Ro, E, M, C.

H. humifusum Trailing St John's-wort; Beachnuadh Làir
Armadale Castle and Dalavil. Once near Balachuirn, Raasay.
Rare. Euro Temp. Skye (24, 33, 50, 60), Ra (54), Ru, E, M.

H. pulchrum Slender St John's-wort; Lus Chaluim Chille
Roadside banks and moorland. Common. SubO Temp. Skye
(all squares), Ra (53-55, 64, 65), Ro, Ru, E, M, C, Sc, So.

H. elodes Marsh St John's-wort; Meas an Tuirc Allta
Old record Loch Gauscavaig, Sleat; Skye (50). Marshy ground
on Canna. Ocea Temp. Probably Canna only.

Drosera rotundifolia Round-leaved Sundew; Lus na Feàrnaich
Wet moorland and bog pools. Common. Circ BoTe. Skye (all squares), Ra (53-55, 64, 65), Ro, Ru, E, M, C, Sc, So.

D. rotundifolia x **D. anglica** = **D**. x **obovata** *
"Sparingly on Raasay". Recorded on Skye (42, 43, 51) Ra (54, 55), Ru.

D. anglica Great Sundew; Lus a'Ghadmainn
Wet bogs. More local. Circ BoMo. Skye (all squares except 26, 36, 37, 47, 54), Ra (53-55, 64, 65), Ro, Ru, E, C, Sc, So.

D. intermedia Oblong-leaved Sundew; Dealt Ruaidhe
Either much less common than *D. anglica*, or under-recorded. SubO Temp. Skye (15, 25, 34, 36, 44, 50, 52, 55, 62, 71, 72), Ro, Ru, E.

Viola riviniana Common Dog-violet: Dail-chuach Coitcheann
Grassland, moorland, woodland, cliff ledges, etc. Common. Euro Temp. Skye (all squares), Ra (53-55, 64, 65), Ro, Ru, E, M, C, Sc, So.

V. canina Heath Dog-violet; Sail-chuach Mointeach
Sandy edge of Loch Fiachanis, Rum 1997, confirming 1930s record. ESib BoTe. Ru.

V. palustris Marsh Violet; Dail-chuach Lèana
Bogs and damp grassland. Locally common. Euro BoTe. Skye (all squares), Ra (53-55, 64, 65), Ro, Ru, E, M, C, Sc, So.

V. tricolor Wild Pansy, Heartsease; Brog na Cubhaig
Arable weed, once locally plentiful. Euro Temp. Skye (14, 15, 24, 25, 34-36, 44, 45, 47, 50, 62, 71, 72), Ra (53), E, Sc.

V. arvensis Field Pansy; Luibh Cridhe
Uncommon in cultivated ground. ESib Temp. Skye (15, 24, 42, 43, 45, 50, 61, 72), E, M, Sc.

Populus tremula Aspen; Critheann
Found singly or in small groups along stream banks, or on rocky ledges. Local. EAsi BoTe. Skye (all squares except 37), Ra (53-55, 64, 65), Ro, Ru, E, M, C, Sc, So.

Salix pentandra Bay Willow; Seileach
Probably always planted and used for basket-making or for shelter. Rare. ESib BoTe. Skye (24, 32, 36, 37, 44, 46, 47), Ru.

S. fragilis Crack-willow; Seileach Brisg *
As above. Rare. ESib Temp. Skye (15, 47, 53, 61, 62), Ru, E.

S. alba White Willow; Seileach Bàn *
As above. All records in NW Skye. Rare. ESib SoTe. Skye (14, 15, 24, 32, 34), Ru, E.

S. triandra Almond Willow
Introduced, N and W Skye. EAsi Temp. Skye (14, 15, 24, 26, 33, 35, 46, 51, 53, 56), Ra (54).

S. purpurea Purple Willow; Seileach Corcorach
Scattered specimens, probably all planted. Commoner in N Skye. ESib Temp. Skye (24, 34, 35, 44-47, 52, 53, 60, 62), E, C, So.

S. viminalis Osier; Seileach Uisge
Often planted in the past for basket-making. EAsi Temp. Skye (15, 24, 26, 32, 34-37, 42-47, 50-53, 56, 60-62, 71, 72), Ra (53-55, 64, 65), Ro, Ru, E, M, C.

S. caprea Goat Willow; Sùileag
Scattered specimens in woodland, and in rocky stream gorges. Commoner in N Skye, rare in Sleat. EAsi BoTe. Skye (14, 24-26, 33-36, 43-47, 51-56, 60-62, 71, 72), Ra (53-55, 64, 65), Ro, Ru, E, C, Sc, So. Ssp. *sphacelata* Skye (42, 56).

S. cinerea ssp. **oleifolia** (ssp. *atrocinerea) Salix atrocinerea* Grey Willow; Seileach Ruadh
Damp woods, stream banks. Locally common and very variable. ESib BoTe. Skye (all squares except 34, 37, 47), Ra (53-55, 64, 65), Ro, Ru, E, M (introduced), Sc, So.

S. aurita Eared Willow; Seileach Cluasach
The commonest willow, growing in heathland and moorland, along streams, and on cliff ledges. Euro BoTe. Skye (all squares), Ra (53-55, 64, 65), Ro, Ru, E, M, C, Sc, So.

S. phylicifolia Tea-leaved Willow; Seileach Thuathach
About two plants to each square! Allt Dearg Beag and Allt Coir' a' Mhadaidh, 42; bank of burn at head of Lon Mor, c.1100ft, 45; Lon Horro, 47. Rare. Circ BAMo. Skye (42, 45, 47, 62, 72), Ru.

S. repens Creeping Willow; Seileach Làir
Moorland, rock ledges on hills. Includes the coastal var. **argentea**. Locally common, except in Sleat. ESib BoTe. Skye (all squares), Ra (53-55, 64, 65), Ro, Ru, E, M, C, Sc, So.

S. myrsinites Whortle-leaved Willow; Seileach Miortail
Two plants on wet rocks below Sgurr Mor, 47. Rare. Euro ArMo. Skye (47).

S. herbacea Dwarf Willow; Seileach Ailpeach
Hill tops and rocky ledges, usually above 1000ft. Ben Edra; Storr; Cuillin; Red Hills; and others. Locally plentiful. Euro ArMo. Skye (14, 15, 23-25, 32, 33, 42-47, 52, 53, 55, 62, 71, 72), Ra (53, 54), Ru, E.

Salix hybrids recorded so far:-

Salix x **sericans**, Ra (53); **S.** x **multinervis** Skye (26, 56), Ra (53), Ru, E; **S.** x **laurina**, Rum; **S.** x **ambigua** (commonest). Skye (47, 72), Ru, E; **S.** x **latifolia** Skye (56); **S** x **reichardtii** Ra (53, 54), Ro.

Sisymbrium officinale Hedge Mustard; Meilise
Waste ground, Kyleakin. Rare. Euro SoTe. Skye (62, 72), Ru.

Alliaria petiolata Garlic Mustard, Hedge Garlic; Gairleach Callaid
Shady gardens. Dunvegan; Corry Lodge, Broadford. Introduced. Euro Temp. Skye (24, 34, 36, 62).

Arabidopsis thaliana Thale Cress; Biolair Thàilianach
Rare on dry rock ledges and in fine scree. Garden weed, Kyleakin and Prabost. ESib Temp. Skye (15, 24, 25, 33, 35-37, 42, 44-47, 50, 51, 54-56, 60-62, 72), Ra (53-55), Ru, M.

Hesperis matronalis Dame's-violet; Feasgar-lus
Garden escape, old road Broadford-Kyleakin. Skye (72).

Barbarea vulgaris Winter-cress; Treabhach
Arable weed, Portree and Carbost (44). Casual, Prabost (45). Introduced. ESib Temp. Skye (44, 45, 62, 72), Ru.

B. intermedia Medium-flowered Winter-cress; Treabhach Meadhanach
Garden weed in Portree. Introduced. Skye (44), Eyre, Ra (53), Ru.

Rorippa nasturtium-aquaticum *Nasturtium officinale* Water-cress; Biolair Uisge
Locally plentiful in streams and springs. ESib SoTe. Skye (14, 15, 24, 25, 33, 35, 37, 41, 43-46, 50-52, 55, 61, 62), Ra (53), E, M, C.

R. microphylla *N. microphyllum* Narrow-fruited Water-cress;
Needs separating from above. Seeds in one row instead of two, in each cell. Skye (36, 60).

Cardamine pratensis Cuckooflower; Lady's-smock; Flùr na Cuthaig
Common in damp places. Circ WiBo. Skye (all squares), Ra (53-55, 64, 65), Ro, Ru, E, M, C, Sc, So.

C. flexuosa Wavy Bitter-cress; Searbh-bhiolair Chasta
Damp shady places. Common. Euro Temp. Skye (all squares except 41), Ra (53-55, 64, 65), Ro, Ru, E, M, C, Sc, So.

C. hirsuta Hairy Bitter-cress; Searbh-bhiolair Ghiobach
Less common, on bare ground, but can become an annoying arable weed. ESib SoTe. Skye (14, 15, 23-25, 32, 34-37, 41, 44-47, 50-52, 54, 55, 60-62, 72), Ra (53-55), Ru, E, M, C, Sc, So.

Arabis petraea *Cardaminopsis petraea* Northern Rock-cress; Biolair na Creige Thuathach
Frequent on rock ledges and among scree, Storr-Quiraing ridge; Cuillin; Blaven. Scarce on Red Hills. Refound on Raasay, 2003. EAsi ArMo. Skye (41-47, 52-55, 62), Ra (53), Ru.

A. alpina Alpine Rock-cress; Biolair na Creige Ailpeach
On wet rock ledges in Coire na Creiche and (2003) Coire a'Bhasteir, Cuillin hills. The only British localities. ESib ArMo. Skye (42).

A. hirsuta Hairy Rock-cress; Biolair na Creige Ghiobach
Scattered plants on rock ledges, not confined to limestone. Rare. Circ BoTe. Skye (24, 26, 33, 36, 45-47, 51-55, 61, 62), Ra (53-55), Ru, E, Sc.

Draba norvegica Rock Whitlowgrass; Biolradh Gruagain an t-Slèibhe
Discovered in 1969 on rock ledges below Sgurr nan Gillean. "Doubtful" 1964 record from Coire na Creiche confirmed 1975. Below Sgurr Thormaid 1980. Blaven 1999. Rare. Euro ArMo. Skye (42, 52).

D. incana Hoary Whitlowgrass; Biolradh Gruagàin Liath
Scattered plants on dry basic rock ledges and in fine scree. Local. Euro BAMo. Skye (14, 15, 25, 26, 32, 33, 36, 42, 44-47, 52, 54, 55), Ra (53, 54), Ru, E.

Erophila verna agg. Common Whitlowgrass; Biolradh Gruagàin
Locally plentiful on grassy knolls grazed by sheep, or on dry paths. Has often flowered in January. ESib SoTe. Skye (24, 25, 33-37, 42, 44, 45, 47, 51, 55, 60, 62, 71, 72), Ra (53, 54), Ru. Ssp. **glabrescens** Skye (45, 62), Ra (53).

Cochlearia officinalis agg. Common Scurvygrass; Carran
Rocks and cliffs by the sea; salt-marshes; mountains. Locally common. Circ WiBo. Skye (all squares), Ra (53-55, 65), Ro, Ru, E, M, C, Sc, So. Within this complex and highly variable aggregate, there are at least two taxa in our area. **C. pyrenaica** ssp. **alpina** (*C. alpina, C. officinalis* ssp. *alpina*): damp cliff ledges in the mountains, Cuillin, Trotternish. Euro ArMo Skye (42, 45, 46).

C. officinalis ssp. **scotica** (*C. scotica*): coastal areas only. Flowers often mauve or lilac. Overlooked? Endemic. Skye (14, 15, 25, 26, 31, 32, 34, 43, 45-47, 50, 52, 60), Ra (53), Ru, M, So.

C. danica Danish Scurvygrass; Carran Danmhairceach
Maritime only. Rare or under-recorded? Rubha Hunish, 47. Ocea Temp. Skye (47, 56, 60, 62), Ra (53), Ru, E, M, Sc.

Capsella bursa-pastoris Shepherd's-purse; Sporan Buachaille
Fairly common arable weed. ESib WiTe. Skye (all squares except 23, 25, 37, 41, 46, 53-56), Ra (53), Ro, Ru, E, M, C, Sc, So.

Thlaspi caerulescens *T. alpestre* Alpine Penny-cress; Praiseach Ailpeach
Basic rocks, Fionchra and Bloodstone, Rum, only. Euro BoMo. Ru.

Lepidium heterophyllum *L. smithii* Smith's Pepperwort; Piobar an Duine Bhochd
Dry banks and road verges. Rare. Ocea SoTe. Skye (33, 42, 45) (in 1950s only), Eigg (1996).

Subularia aquatica Awlwort; Lus a'Mhinidh
Easily overlooked in base-poor lochs and pools. Local. Circ BoMo. Skye (14, 32-34, 41-44, 46, 50, 60-62, 72), Ru, Sc.

Brassica napus Rape; and Turnip or Swede; Raib
Sometimes grown as fodder crop - probably casual only. Introduced. Skye (24, 26, 34), Ra (53, 54), E, M.

Sinapis arvensis Charlock; Sgeallan
Arable weed, once locally common. ESib Temp. Skye (24, 26, 35, 36, 47, 50, 52, 60, 62, 72), Ra (53, 54), Ru, E, M, C.

Cakile maritima Sea Rocket; Fearsaideag
Greep; Glen Brittle beach; Camas Ban, Portree. Rare due to lack of suitable habitat on Skye. Euro WiTe. Skye (24, 42, 44), Ru, E.

Raphanus raphanistrum ssp. **raphanistrum** Wild Radish; Meacan Ruadh Fiadhain
Arable weed, once widespread but not common. Introduced. Euro SoTe. Skye (14, 15, 24-26, 33, 34, 42-45, 50, 52, 60, 62, 71), Ra (53), Ru, E, C.

Empetrum nigrum ssp. **nigrum** *E. nigrum* Crowberry; Lus na Feannaig
Moorland; often on sea cliffs. Locally common. Circ BAMo. Skye (all squares), Ra (53-55, 64, 65), Ro, Ru, E, M, C, Sc, So.

E. nigrum ssp. **hermaphroditum** *E. hermaphroditum* Mountain Crowberry; Dearcag Fithich

Usually on higher ground. Young stems green, not red. Further work required. Circ BAMo. Skye (24, 42, 43, 46, 47, 51-53, 72), Ru.

Rhododendron ponticum Rhododendron

Introduced and naturalised near original plantings - Dunvegan, Kilmarie, Armadale. Local. Skye (24, 25, 51, 60, 62, 71), Ra (53, 54), Ru, E, C, So.

Loiseleuria procumbens Trailing Azalea; Lusan Albannach

Old records from the 1880s refound in small quantities on Cuillin Hills, 1981-82; Kyleakin Hills, 1986. Circ ArMo. Skye (42, 71, 72).

Gaultheria mucronata Prickly Heath

Naturalised on cliffs and roadsides around Kilmarie estate, Strathaird. Introduced. Skye (51), Ra (53), Ru.

Arctostaphylos uva-ursi Bearberry; Grainnseag

Moors, rock faces. Locally common in S Skye. Isolated patches in N Skye. Circ BoMo. Skye (14, 15, 24-26, 32, 35, 36, 41-43, 46, 47, 50-53, 55, 60-62, 71, 72), Ra (53-55, 64, 65), Ro, Ru, Sc, So.

A. alpinus Arctic Bearberry; Grainnseag Dubh

Recorded from 'Beinn na Greine' by Lightfoot, 1772. Refound on Beinn Bhuidhe (same area) 1990. Circ ArMo. Skye (72).

Calluna vulgaris Heather, Ling; Fraoch

Growth and area covered affected everywhere by uncontrolled burning and by sheep. Common. Euro BoTe. Skye (all squares), Ra (53-55, 64, 65), Ro, Ru, E, M, C, Sc, So.

Erica tetralix Cross-leaved Heath; Fraoch Frangach

Wet moors and bogs. Common. SubO Temp. Skye (all squares), Ra (53-55, 64, 65), Ro, Ru, E, M, C, Sc, So.

E. cinerea Bell Heather; Biadh na Circe Fraoich

Drier moors, and so is also over-grazed and burned. Common. Ocea Temp. Skye (all squares), Ra (53-55, 64, 65), Ro, Ru, E, M, C, Sc, So.

Vaccinium microcarpum Small Cranberry; Muileag

On *Sphagnum*, boggy ground, Wiay, and edge of lochan in Sleat. Circ BoMo. Skye (23, 61).

V. vitis-idaea Cowberry; Lus na Braoileag
Usually over 1000 ft on Storr-Quiraing hills; Cuillin; Sgurr na Coinnich; etc. Never more than 2" high, except on sheltered cliffs. Occasional. Circ BAMo. Skye (14, 23-25, 32-35, 42-47, 50, 52, 53, 55, 60-62, 71, 72), Ra (53, 54), Ru, E, Sc.

V. vitis-idaea x **V. myrtillus** = **V.** x **intermedium**
Plants which are evergreen and puberulent, two characteristics of the hybrid, have been found all over Skye in winter, often miles from the nearest *V. vitis-idaea.* Until one with either flower or fruit (again distinctive for the hybrid) can be found, their identity is uncertain. Cavalôt (2004 *BSBI News* 96: 18-22) discusses the problems of identification in detail.

V. myrtillus Blaeberry, Bilberry; Dearca-fFraoich
Moorland, rock ledges in stream gorges. Common, though fruit is not. ESib BoMo. Skye (all squares), Ra (53-55, 64, 65), Ro, Ru, E, Sc, So.

Pyrola minor Common Wintergreen; Glas-luibh Beag
Among heather at Brae, Raasay, 1969; also recorded from Screapadal. Third site on Raasay, 1993. Dunvegan Castle woods, 1975. Hazel scrub, Skerinish, 1999. Rare. Circ BoMo. Skye (24, 45), Ra (53, 54), Ru.

P. media Intermediate Wintergreen; Glas-luibh Meadhanach
Usually on thin soil, below heather. In the open at 380 ft on Ben Tote (44), and there are a dozen other moorland sites within a 3-mile radius. Also in Glendale and Sleat (one site each). Beinn Bhreac 1985; Raasay 1996. Eigg record (1930s) never refound. Cannot be separated from above when not in flower. Very local. ESib BoMo. Skye (14, 25, 34, 35, 44, 45, 61), Ra (54).

P. rotundifolia ssp. **rotundifolia** Round-leaved Wintergreen; Glas-luibh Cruinn
E. coast cliff on Raasay 1997. Rare. ESib BoTe. Ra (54).

Orthilia secunda Serrated Wintergreen; Glas-luibh Fhiaclach
With *P. minor* at Brae, Raasay 1969, and in two other sites in the same square; cliffs above Fearns, 1994. 'Sligachan' record of 1868 never refound, but several groups discovered in wooded gorge, Glen Arroch, 1978. Rare. Circ BoMo. Skye (72), Ra (53, 54).

Primula vulgaris Primrose; Sobhrag
Open grassland, woodland banks, rock ledges on hills. Common. Euro Temp. Skye (all squares), Ra (53-55, 64, 65), Ro, Ru, E, M, C, Sc, So.

Lysimachia nemorum Yellow Pimpernel

Wet places. Locally plentiful. SubO Temp. Skye (all squares), Ra (53, 54, 65), Ro, Ru, E, M, C, Sc, So.

Anagallis tenella Bog Pimpernel; Falcair Lèana

Tiny isolated patches (compared with the Outer Isles), commoner in S Skye. Rare. Ocea SoTe. Skye (14, 32, 41, 42, 45, 47, 50, 61), Ru, E, M, C, Sc.

A. arvensis Scarlet Pimpernel; Falcair

Arable weed. Near Clachan, Raasay (1930s). Casual, Prabost, 1990 and 1998. Rare with decline of arable agriculture. ESib SoTe. Skye (36, 45, 62), Ra (53), E, M, C.

A. minima *Centunculus minimus* Chaffweed; Falcair Mìn

Bare, damp ground near the sea, or muddy paths inland. Rare. Euro Temp. Skye (14, 24, 25, 33, 47, 60, 61, 72), Ra (53, 54, 64), Ro, Ru, E, C, Sc, So.

Glaux maritima Sea-milkwort; Lus na Saillteachd

Salt-marshes or rocky shores. Common. Circ BoTe. Skye (all squares except 41, 54, 55), Ra (53-55, 65), Ro, Ru, E, M, C, Sc, So.

Samolus valerandi Brookweed; Luibh an t-Sruthain

Wet rocky ground near the sea. Aird and Tormore, Sleat; mouth of Allt na Leac, S of Camas Malag; also 41. Rare. Circ SoTe. Skye (41, 50, 51, 60).

Ribes spicatum Downy Currant; Raosar Giobach

One or two bushes in boulder scree, Sgurr a'Bhagh, 25; Kingsburgh, 35; at least 100 plants below Waterstein Head, in scree, 14. Rare. Euro BoTe. Skye (14, 25, 35).

Ribes nigrum Black Currant and **Ribes uva-crispa** Gooseberry sometimes occur as garden escapes.

Crassula helmsii New Zealand Pigmyweed

Introduced in 1980s to man-made ponds, Skeabost House Hotel and Armadale Castle. Skye (44, 60).

Sedum rosea *Rhodiola rosea* Roseroot; Lus nan Laoch

On sea cliffs and on rock ledges on hills. Locally plentiful. Circ ArMo. Skye (all squares except 34), Ra (53-55, 64, 65), Ro, Ru, E, M, C, Sc, So.

S. acre Biting Stonecrop, Wall-pepper; Grabhan nan Clach

Sandy grassland and shingle. Stonework of Armadale pier; edge of shore at Fiskavaig; Glen Brittle; Camasunary. Very local. Euro Temp. Skye (25, 33, 36, 42, 46, 47, 50-52, 60, 61, 71), Ra (53), Ru, E, M, C, So.

S. anglicum English Stonecrop; Biadh an-t-Sionnaidh
Rocks or stony places near the sea, and on hills. Local. Ocea Temp. Skye (all squares), Ra (53-55, 64, 65), Ro, Ru, E, M, C, Sc, So.

Saxifraga nivalis Alpine Saxifrage; Clach-bhriseach an t-Sneachda
Rare on rock ledges and among boulders, Storr and Quiraing areas, and Fionchra, Rum. Even rarer on damp ledges high in Cuillin corries. Circ ArMo. Skye (42, 45, 46), Ru.

S. stellaris Starry Saxifrage; Clach-bhriseach Reultach
Locally common, springs and wet ledges. Storr-Quiraing hills, Red Hills, Cuillin and others. Euro ArMo. Skye (24, 32, 33, 42-47, 51-53, 55, 71, 72), Ru.

S. oppositifolia Purple Saxifrage; Clach-bhriseach Phurpaidh
Locally common on rock ledges and stony ground, Storr-Quiraing hills; Blaven (also at sea level, head of Loch Slapin). Rare on Healaval Bheag and the Cuillin. Missing from Raasay. Circ ArMo. Skye (14, 15, 24-26, 41, 42, 44-47, 51-56), Ru, E.

S. aizoides Yellow Saxifrage; Clach-bhriseach Bhuidhe
Wet sea cliffs; mountain scree, basic cliffs and river gravel. Local. Euro ArMo. Skye (24, 36, 45-47, 51-56, 61, 62, 71, 72), Ra (53, 54), Ru, E.

S. hypnoides Mossy Saxifrage; Clach-bhriseach Còinnich
Dunvegan Head; Storr-Quiraing hills; Blaven group; etc. Not yet recorded from Cuillin. Locally plentiful. Ocea BoMo. Skye (14, 15, 24, 25, 32, 33, 36, 44-47, 52-55, 71), Ra (53, 54), Ru, E, C.

Chrysosplenium oppositifolium Opposite-leaved Golden-saxifrage; Lus nan Laogh
Wet rocks, beside springs, or in marshy ground. Common. SubO Temp. Skye (all squares), Ra (53-55, 64, 65), Ro, Ru, E, M, C, Sc, So.

Parnassia palustris Grass-of-Parnassus; Fionnan Geal
Most records from Cambrian limestone (Suardal, Torrin, Tokavaig) or Jurassic rocks (E side of Trotternish). Local. Circ BoTe. Skye (33, 44-46, 50-52, 54-56, 60-62, 72), Raasay record (1930s) never refound. Ru, E, M, C.

Spiraea salicifolia Bridewort, Willow Spiraea
Garden escape. Introduced. Skye (36, 44, 50, 51, 61, 62, 71, 72), Ru.

Filipendula ulmaria Meadowsweet; Lus Chuchulainn

Wet grassland, ditches. Common. EAsi BoTe. Skye (all squares), Ra (53-55, 64, 65), Ro, Ru, E, M, C, Sc, So.

Rubus saxatilis Stone Bramble; Caor Bad Miann

Steep banks, rock ledges on moors and hills. Local. EAsi BoTe. Skye (all squares except 37, 56), Ra (53, 54, 64, 65), Ru, E, C, Sc, So.

R. idaeus Raspberry; Suibheag

Edges of woods and grassy banks. Locally plentiful. Circ BoTe. Skye (all squares except 23, 37, 41), Ra (53-55, 64, 65), Ro, Ru, E, M, C, Sc, So.

R. fruticosus agg. Bramble, Blackberry; Dris (bush) / Smeur (berry)

Woodland, scrub, roadside banks and verges. Local. Commoner in S. Skye, and in woodland at Dunvegan. Euro SoTe. Agg. Skye (all squares except 23, 37, 56), Ra (53, 54, 65), Ro, Ru, E, M, C, Sc, So. Careful investigations of microspecies essential: the following have been confirmed by experts:

R. dasyphyllus Ru.

R. dumnoniensis Ru, E.

R. ebudensis Skye (24, 35).

R. hebridensis Skye (24, 35).

R. laciniatus

Garden escape near Orbost, and Scorrybreck, Portree (now gone). Introduced. Skye (24, 44), M.

R. latifolius Skye (51, 62).

R. leptothyrsos Skye (62).

R. lindleianus Skye (44, 60, 71, 72).

R. mucronulatus Skye (24, 72), Ru, So.

R. nemoralis Skye (24, 35, 44, 52, 60-62, 71, 72), Ru, E, So.

R. plicatus Skye (44).

R. polyanthemus Skye (24, 41, 42, 46, 52, 60-62, 71, 72), E, M.

R. pyramidalis Skye (72).

R. radula Skye (24, 25, 35, 36, 44, 45, 72), Ra (53), E.

R. septentrionalis Skye (24, 35, 60, 62, 71), Ru.

R. spectabilis Introduced. Skye (24, 60, 62).

R. sprengelii Ru.

R. subinermoides Skye (44, 62).

R. venetorum E.

Potentilla palustris *Comarum palustre* Marsh Cinquefoil; Coig-bhileach Uisge
Marshy ground, and wet margins of lochs. Locally plentiful. Circ BoTe. Skye (all squares except 41, 53, 54), Ra (53, 54), Ro, Ru, E, M, C, Sc, So.

P. anserina Silverweed; Brisgean
Waste ground, shingle, and sandy grassland. Common. Circ BoTe. Skye (all squares), Ra (53-55, 64, 65), Ro, Ru, E, M, C, Sc, So.

P. crantzii Alpine Cinquefoil; Leamhnach Ailpeach
Short turf and rock ledges. Ben Suardal, 1981. Old record from Rum never confirmed. ESib BAMo. Skye (62).

P. erecta ssp. **erecta** Tormentil; Cairt-leamhna
Very common on moorland, hill grassland, and open woodland. ESib BoTe. Skye (all squares), Ra (53-55, 64, 65), Ro, Ru, E, M, C, Sc, So. Ssp. **strictissima** has been recorded on Skye (52, 61, 62), and probably occurs in more squares.

P. reptans Creeping Cinquefoil; Coig-bhileach
Marble quarry road, Suardal, 62. Probably introduced. ESib SoTe. Old records in 24, and 56 unconfirmed. Skye (62).

P. sterilis Barren Strawberry; Subh-làir Brèige
Noticeable in dry grassy places in the spring; not so obvious later on. SubO Temp. Skye (all squares except 15, 36, 37, 55), Ra (53, 55), Ru, E.

Sibbaldia procumbens Sibbaldia; Siobaldag
Grassland and scree on Trotternish ridge between Ben Edra and Bealach Chaiplin, 46. Also below Sgurr Mor, 47, Blaven, 52, and Sgurr na Coinnich, 72 (1983). Doubtful Storr and Cullin records need confirmation. Circ ArMo. Skye (46, 47, 52, 72).

Fragaria vesca Wild Strawberry; Subh-lair
Dry grassy banks and rock ledges, most plentiful on ultra-basic and limestone. Local. ESib Temp. Skye (all squares except 37), Ra (53-55, 64, 65), Ro, Ru, E, M, C, Sc, So.

Geum rivale Water Avens; Machall Uisge
Wet grassland, ditches, and wet rock ledges on hills. Local. ESib BoTe. Skye (all squares), Ra (53-55, 64, 65), Ro, Ru, E, C, Sc, So.

G. rivale x **G. urbanum** = **G.** x **intermedium** Hybrid Avens
Wood S of Portree, 44; also 60 (Skye) and perhaps elsewhere, if parents occur together. Skye (44, 60), E.

G. urbanum Wood Avens; Machall Coille
Restricted to areas with woodland or scrub to provide shade.
ESib Temp. Skye (14, 24, 34-36, 41, 42, 44, 47, 50, 54, 60-62,
71, 72), Ra (53, 54, 64), Ru, E, C.

Dryas octopetala Mountain Avens; Machall Monaidh
Locally plantiful on limestone at Suardal, and Camas Malag, S
of Torrin. Also scattered patches on Dunvegan Head and Ben
Tianavaig (both basalt); and E side of Raasay (limestone). Circ
ArMo. Skye (15, 47, 51, 54, 55, 61, 62), Ra (53, 54), Ru, E.

Agrimonia eupatoria Agrimony; Geur-bhileach
Occasional plants in grassland, usually near the sea. ESib
SoTe. Skye (24, 25, 33, 36, 41, 44, 45, 51, 53-55, 61), Ra (53),
E.

A. procera Fragrant Agrimony; Geur-bhileach Chubhraidh
Woodland, Portree; N. of Hallaig, Raasay. Rare. Euro Temp.
Skye (44), Ra (54).

Acaena inermis *Acaena microphylla* Spineless Acaena, New
Zealand Bur
Naturalised near the pier and along the old railway track,
Raasay. Introduced. Ra (53).

Alchemilla alpina Alpine Lady's-mantle; Trusgan
Mountain grassland, rocks and scree; descends to sea level at
Torrin, 52. Locally common. Euro ArMo. Skye (all squares
except 14, 15, 23, 34, 35, 37, 50, 60), Ra (53, 54), Ru, C, Sc.

A. xanthochlora Lady's Mantle; Fallaing Moire Bhuidhe
Grassland at lower levels. Possibly under-recorded. Still
missing from Raasay. Euro Temp. Skye (all squares except 32,
37, 41, 50, 53), Ru, E, So.

A. filicaulis ssp. **filicaulis** Lady's Mantle; Fallaing Moire Chaol
Hill and mountain grassland. Less common than ssp. **vestita**.
Euro BoMo. Skye (15, 25, 34, 36, 45, 46, 53, 54, 62, 71, 72),
Ru.

A. filicaulis ssp. **vestita** Lady's Mantle; Fallaing Moire Chaol
Basic grassland and hills. Local. Euro BoMo. Skye (15, 24, 25,
33, 35, 36, 45-47, 51-55, 61, 62, 71), Ra (53-55, 65), Ru, E,
Sc, So.

A. wichurae Lady's Mantle; Fallaing Moire Ailpeach
Recorded from 3 places in the Storr area, on wet rock ledges.
Rare. Euro BoMo. Skye (45, 55).

A. glabra Lady's Mantle; Fallaing Moire Mhin
Grassland, rock ledges. Common. Euro BoTe. Skye (all
squares), Ra (53, 54, 65), Ru, E, M, C, So.

Aphanes arvensis Parsley-piert: Spionan Moire
Weed of cultivated and waste ground. Probably under-recorded, as *A. arvensis* is difficult to separate from *A. australis* when not in fruit. EuroTemp. Skye (24, 25, 34-36, 44, 46, 60, 62, 71), Ra (53-55), Ro, Ru, E, C, Sc.

A. australis *A. microcarpa* Slender Parsley-piert; Spionan Moire Caol
Arable weed on croftland, Prabost, 45; and elsewhere. Euro Temp. Skye (14, 24, 26, 33, 35, 45, 51, 60, 61, 72), Ra (53, 54), Ro, E, M.

Rosa pimpinellifolia *R. spinosissima* Burnet Rose; Ròs Beag Bàn na h-Alba
Rough grassland, rock ledges; most plentiful on limestone. Local. Rare in Trotternish north of Portree; missing from Raasay. EAsi Temp. Skye (13-15, 23-25, 32, 33, 35, 36, 41-44, 50-53, 60-62), Ru, E, M, C, Sc, So.

R. rugosa Japanese Rose; Ròs Rocach
Naturalised in several places, including shingle at Bornaskitaig 37. Introduced. Skye (14, 26, 36, 37, 44, 45, 50, 52), Ra (53), M, So.

R. canina Dog-rose; Ròs nan Con
Scrub woodland, grassy banks. Possibly under-recorded. Euro Temp. Skye (14, 15, 23-26, 33-36, 41, 42, 50-52, 54, 55, 60-62, 71, 72), Ra (53-55, 64, 65), Ro, Ru, E, M, C, Sc, So.

R. caesia Glaucous Dog-rose; Ròs nan Con Tuathach
Similar habitats to *R. canina*. Under-recorded. Euro Temp. Skye (36, 44, 45), Ra (54), Ru, E.

R. sherardii Sherard's Downy-rose; Ròs Shioraird
Similar habitats to above. More work required. Euro Temp. Skye (25, 26, 42-46, 50, 52, 55, 61, 71), Ra (53-55, 64, 65), Ro, Ru, E, M, C, Sc, So.

R. mollis Soft Downy-rose; Ròs Bog
Probably commoner than records suggest. Euro BoTe. Skye (14, 15, 24, 25, 33, 34, 36, 44, 45, 50-52, 54, 56, 61, 62, 71, 72), Ra (53, 54), Ru, E, M, C, Sc, So.

R. rubiginosa Sweet-briar; Dris Chubhraidh *
Introduced? Skye (47, 50, 72), Ru.

Hybrids *R. canina* x *R. mollis* (Skye (71)), *R. mollis* x *R. rubiginosa* (Skye (52)), *R. mollis* x *R. pimpinellifolia* (E), and *R. caesia* x *R. canina* (Ra (53), E) have been recorded and identified by experts. Also plants in *R. canina* 'Dumales' (Skye (72), E) and *R. canina* group 'Transitoriae' (E). More work required!

Prunus spinosa Blackthorn; Preas nan Airneag

Probably always planted, as it is at Flodigarry, Duisdale, and in Raasay. Rare. Euro Temp. Skye (25, 47, 51, 60, 71), Ra (53), Ru, E, M, So.

P. padus Bird Cherry; Fiodhag

Occurs as single specimens in woodland and gorges. Rare. EAsi BoTe. Skye (24, 26, 32, 34, 35, 41-46, 50-53, 60-62, 71, 72), Ru, E, Sc.

Sorbus aucuparia Rowan; Caorann

Flourishes on rock ledges out of reach of sheep; more stunted where it has been eaten. EAsi BoTe. Skye (all squares), Ra (53-55, 64, 65), Ro, Ru, E, M, C, Sc, So.

S. rupicola Whitebeam; Gall-uinnsean na Creige

On basic or limestone cliffs, Suishnish, Rudha na Leac, Raasay. In woodland at sea level, Drinan. Usually as single small trees. Rare. SubO BoMo. Skye (51), Ra (53), Ru.

Cotoneaster integrifolius Entire-leaved Cotoneaster

Naturalised on cliff (Jurassic) at Scorrybreck, Portree; also at Camas Malag, S of Torrin, and Greshornish. Planted elsewhere. Introduced. Skye (35, 44, 51, 61), Ru, E.

C. simonsii Himalayan Cotoneaster

Naturalised on cliffs S of Kilmarie 51. Introduced. Skye (25, 45, 51, 61, 71), Ra (53, 54), Ru, E.

Crataegus monogyna Hawthorn; Sgitheach

Commoner in areas with some woodland. Euro Temp. Skye (all squares except 37, 55, 56), Ra (53, 54), Ru, E, M, C.

Anthyllis vulneraria Kidney Vetch; Cas an Uain

Road verges, moorland, seaside, and mountain cliff-ledges. Local. Euro BoTe. Skye (all squares except 32, 53, 71), Ra (53-55), Ru, E, M, C. Ssp. **lapponica** in Skye (14, 44, 54, 55), Ru.

Lotus corniculatus Common Bird's-foot-trefoil; Peasair a'Mhadaidh Ruadh

Roadsides, grassy places, moors. Common. EAsi SoTe. Skye (all squares), Ra (53-55, 64, 65), Ro, Ru, E, M, C, Sc, So.

L. pedunculatus *L. uliginosus* Greater Bird's-foot-trefoil; Barra-mhislean Leana

Damp places. Very local. Euro Temp. Skye (14, 15, 24, 25, 33-36, 42, 44, 45, 51-52, 56, 60-62, 71, 72), Ra (53, 54), Ru, E, C.

Vicia orobus Wood Bitter-vetch; Peasair Shearbh
> On rock and cliff ledges in NW Skye; roadside banks at Roag (24) and Aird, Sleat (60). Rare. SubO Temp. Skye (14, 15, 23-26, 31, 35, 60), Ru, M, C.

V. cracca Tufted Vetch; Peasair nan Luch
> Common in grassy places, and as an arable weed. EAsi BoTe. Skye (all squares except 41), Ra (53, 54, 64), Ru, E, M, C, Sc, So.

V. sylvatica Wood Vetch; Peasair Coille
> Most records so far from cliffs and gullies near the sea. Missing from Raasay. Rare. ESib BoTe. Skye (14, 24-26, 41, 51, 55, 56, 61), Ru, E, C.

V. hirsuta Hairy Tare; Peasair an Arbhair
> Casual at Prabost, 1957; wasteground, Kyleakin, 1978. Introduced? Euro Temp. Skye (24, 25, 45, 51, 62, 72), Ra (53), E (old record), C.

V. sepium Bush Vetch; Peasair nam Preas
> Grassy places and rock ledges. Common. ESib BoTe. Skye (all squares), Ra (53-55), Ro, Ru, E, M, C, Sc, So.

V. sativa ssp. **nigra** *V. angustifolia* Common Vetch; Peasair nan Coilleag
> Gardens, road verges. Rare. Introduced? Euro SoTe. Skye (52, 60, 62, 71, 72), Ra (53), E, C.

V. sativa ssp. **sativa** Common Vetch; Peasair nan Coilleag *
> Weed of clutivated ground. Very rare, no recent records. Possibly overlooked. Skye (45-47).

Lathyrus linifolius *L. montanus* Bitter-vetch; Carrachan
> Grassy banks and rock ledges - the first vetch to flower. Common. Euro Temp. Skye (all squares), Ra (53, 54), Ru, E, M, C, Sc, So.

L. pratensis Meadow Vetchling; Peasair Bhuidhe
> Arable grassland. Common. ESib BoTe. Skye (all squares except 41), Ra (53-55), Ro, Ru, E, M, C, Sc, So.

Medicago lupulina Black Medick, Dubh-mheidig *
> Grassland. Old records - casual only? ESib Temp. Skye (36, 52), Ra (53), Ru, M, C.

Trifolium repens White Clover; Deochdan Geal
> Arable grassland. Common. ESib BoTe. Skye (all squares), Ra (53-55, 64, 65), Ro, Ru, E, M, C, Sc, So.

T. hybridum Alsike Clover; Seamrag Shuaineach
> Weed of cultivation, Muck 2000. Introduced. Ra (53), Ru, M.

T. campestre Hop Trefoil; Seamrag Bhuidhe
Sandy grassland, Laig, Eigg 1998. ESib SoTe. E, M (old record), C (old record).

T. dubium Lesser Trefoil; Seangan
Roadsides and grassy places. Occasional. Euro Temp. Skye (all squares except 23, 32, 37, 41, 52, 53), Ra (53), Ru, E, M, C, So.

T. pratense Red Clover; Deochdan Dearg
Grassy places, and often sown in hay mixtures. Common. ESib Temp. Skye (all squares), Ra (53-55, 64, 65), Ro, Ru, E, M, C, Sc, So.

T. medium Zigzag Clover; Seamrag Chro-dearg
Grassy places, locally common. ESib BoTe. Skye (14, 15, 24-26, 36, 37, 45-47, 50-52, 56, 60, 62), Ru, E, M, C.

Cytisus scoparius *Sarothamnus scoparius* Broom; Bealaidh
Widely scattered, often solitary plants, usually near houses. Introduced? Euro Temp. Skye (15, 24, 25, 34-36, 42-44, 47, 52, 53, 60-62, 71, 72), Ra (53, 54), Ru, C, Sc.

Ulex europaeus Gorse, Whin; Conas
Rough grassy places, perhaps sometimes planted. Local. Ocea Temp. Skye (all squares except 23, 41, 46, 50, 54, 55), Ra (53, 65), Ro, Ru, E, C, Sc.

U. gallii Western Gorse
Waste ground N. of Portree 1995 but bull-dozed 2002; road verge, Loch Ainort 1997. Introduced. Ocea Temp. Skye (24, 26, 44, 52).

Myriophyllum alterniflorum Alternate Water-milfoil; Snàthainn Bhàthaidh
Prefers base-poor and peaty water in lochs and slowly flowing streams. Locally plentiful. SubO BoTe. Skye (14, 23-26, 32-35, 37, 41-47, 50-53, 56, 60-62, 72), Ra (53, 54), Ro, Ru, E, M, Sc, So.

Lythrum salicaria Purple-loosestrife; Lus na Sìochaint
Isolated patches in marshy ground, Roag and Harlosh; Portree; Braes; Torran, Raasay. Garden escape, Rum. Rare. EAsi Temp. Skye (24, 44, 53), Ra (54, 55), Ro, Ru (escape), E.

Epilobium hirsutum Great Willowherb; Seileachan Mòr
Waste ground, Portree 1997. EAsi SoTe. Skye (44).

E. parviflorum Hoary Willowherb; Seileachan Liath
Rare in marshy ground. Euro Temp. Skye (42, 47, 55, 61), Ra (53), Ro, Ru, C.

E. montanum Broad-leaved Willowherb; Seileachan Coitcheann
Woodland, waste land, and rock ledges on hills. Common.
Euro Temp. Skye (all squares), Ra (53-55, 64, 65), Ro, Ru, E,
M, C, Sc, So.

E. obscurum Short-fruited Willowherb; Seileachan Faireagach
Stream banks, moist ground and rocks. Local. Euro Temp.
Skye (all squares except 23, 34, 41), Ra (53-55, 65), Ro, Ru, E,
M, C, Sc, So.

E. ciliatum *E. adenocaulon* American Willowherb; Seileachan
Aimeireaganach
Waste ground near Druimfearn, 1978 (gone). Uig and Portree
added since. Rare or overlooked? Introduced. Skye (36, 37, 43,
44, 61).

E. palustre Marsh Willowherb; Seileachan Lèana
Marshes, ditches, and bogs. Common. Circ BoTe. Skye (all
squares), Ra (53-55, 64, 65), Ro, Ru, E, M, C, Sc, So.

E. anagallidifolium Alpine Willowherb; Seileachan Ailpeach
Streams and springs, Storr-Quiraing hills; Ben Aslak, Sleat.
Rare. Circ ArMo. Skye (45-47, 71).

E. alsinifolium Chickweed Willowherb; Seileachan Fliodhach
Streams, springs, or moist ledges in Storr-Quiraing hills;
Cuillin; Blaven. Old records from Raasay and Eigg. Rare. Euro
ArMo. Skye (41, 42, 44-47, 52, 55), Ra (54), E.

E. brunnescens *E. nerteroides* New Zealand Willowherb;
Seileachan Làir
Found on paths (Portree, and steps to Storr Lochs generating
station); in quarries (Skerinish and Kyleakin); in scree on hills
(Healaval Mhor, Belig). Local. Introduced. Skye (14, 23-25, 32,
34, 35, 42-47, 51-55, 60, 61, 71, 72), Ra (53-55, 64), Ro, E, M.

Chamerion angustifolium *Epilobium angustifolium, Chamaenerion
angustifolium* Rosebay Willowherb; Seileachan Frangach
In woodland - Dunvegan, Portree, Armadale. Also on sea cliffs.
Locally plentiful. Circ BoTe. Skye (14, 15, 24-26, 32, 33,
35-37, 42-44, 47, 50, 51, 54-56, 60, 72), Ra (53, 54), Ru, E, M,
C, Sc, So.

Fuchsia magellanica Fuchsia
Introduced as a hedge plant, and naturalised in places. Skye
(24-26, 36, 44, 50, 51, 56, 60-62, 72), Ra (53, 54, 65), Ru, E,
M, C.

Circaea lutetiana Enchanter's-nightshade; Fuinseagach

In shade of rock crevices (limestone) at An Leac, 41; and Lusa, 62. Estate woodland - Lynedale, Kilmarie, and Raasay; Port Maol, Muck. Locally plentiful. Euro Temp. Skye (35, 36, 41, 44, 51, 52, 60, 62, 71, 72), Ra (53), Ru, E, M, C, Sc, So.

C. x **intermedia = C. alpina** x **C. lutetiana** Upland Enchanter's-nightshade

Usually in shade of trees or rocks, but grows in shingle in several places. Commoner than above. Skye (24, 25, 32-34, 36, 43-46, 50-56, 60-62, 71, 72), Ra (53, 54, 64), E, Sc.

Ilex aquifolium Holly; Cuilionn

Found wild on rock ledges, in woods, and along rocky stream sides. Local, commoner in S Skye. Also planted. SubO SoTe. Skye (24, 34-36, 41-43, 47, 50-54, 60-62, 71, 72), Ra (53-55, 64, 65), Ro, Ru, Sc, So.

Mercurialis perennis Dog's Mercury; Lus-ghlinne Bhracadail

Dunvegan Castle grounds; Glen Bracadale; St John's Chapel burn; also near Portree, and near Tormore, Sleat; unlocalised record in 62. Very local. Euro Temp. Skye (24, 33, 34, 44, 60, 62), E.

Euphorbia helioscopia Sun Spurge; Lus nan Fionneachan

A weed of cultivated ground, but never in any quantity. EAsi SoTe. Skye (14, 15, 24-26, 34-36, 44, 45, 51-53, 62, 72), Ra (53, 54), Ru, E, M, C.

E. peplus Petty Spurge; Lus Leighis

Garden weed, Kyleakin, 72.; unconfirmed record 47. Rare. Euro SoTe. Skye (72).

Linum catharticum Fairy Flax; Lion nam Ban-sìdh

Grassland, moorland, and rock ledges on hills. Common. Euro Temp. Skye (all squares), Ra (53-55, 64), Ro, Ru, E, M, C, Sc, So.

Polygala vulgaris Common Milkwort; Lus a'Bhainne

Leaves alternate, plant larger and less common than *P. serpyllifolia*, often on basic soils. Euro Temp. Skye (all squares), Ra (53-55, 64), Ru, E, M, C.

P. serpyllifolia Heath Milkwort; Siabann nam Ban-sìdh

Lower leaves opposite. Pink and white forms occur as well as dark and light blue. SubO Temp. Skye (all squares), Ra (53-55, 64, 65), Ro, Ru, E, M, C, Sc, So.

Oxalis acetosella Wood-sorrel; Seamrag

Grassland, woodland, and among rocks and boulders on hills. Reaches summit of Storr (2360') 45, and c. 2700' in Coire na Creiche, 42. EAsi BoTe. Skye (all squares), Ra (53-55, 64, 65), Ro, Ru, E, M, C, Sc, So.

Geranium dissectum Cut-leaved Crane's-bill; Crobh Preachain Giobach

Occasional in cultivated and waste ground. Euro SoTe. Skye (24, 34-36, 44, 50-53, 60-62, 71, 72), Ra (53, 54), Ro, Ru, M, C, So.

G. molle Dove's-foot Crane's-bill; Crobh Preachain Mìn

Occasional in cultivated or sandy ground. Euro SoTe. Skye (24, 33-36, 44, 45, 50, 51, 53, 60-62, 71, 72), Ra (53, 54), Ru, E, M, C, Sc, So.

G. lucidum Shining Crane's-bill; Crobh Preachain Deàlrach

Among old wall debris, Scorrybreck, Portree; walls in Torrin (now gone) and Broadford; Raasay. Rare. SubM SubA. Skye (52, 54, 62), Ra (53, 54).

G. robertianum Herb-Robert; Lus-ros

Woods, rock ledges, scree. Local. Euro Temp. Skye (all squares), Ra (53, 54, 64, 65), Ru, E, M, C, Sc, So.

Erodium cicutarium Common Stork's-bill; Gob Corra

Arable fields or waste places on sandy soil. Dun Scaich, Sleat (1985). ESib SoTe. Skye (51), Ru, E (not since 1940s), M, C. Old records untraced for Skye (26, 36).

Hedera helix Ivy; Eidheann

In woods and on rock faces. Local. Euro SoTe. Skye (all squares), Ra (53-55, 64, 65), Ro, Ru, E, M, C, Sc, So.

Hydrocotyle vulgaris Marsh Pennywort; Lus na Peighinn

Damp ground and ditches, usually near the sea. Very local. SubO SoTe. Skye (23-25, 33, 35-37, 45-47, 50-52, 60, 61, 71), Ra (53-55, 64, 65), Ro, Ru, E, M, C, So.

Sanicula europaea Sanicle; Bodan Coille

In woods, in shade of rocks or boulders, and in limestone grikes. Local. Euro Temp. Skye (all squares), Ra (53-55, 64, 65), Ro, Ru, E, Sc, So.

Astrantia major Astrantia

Garden escape on bank of burn at Camas Mor, 1960s, 37 (Skye). Kinloch woods, Rum. Skye (37), Ru.

Anthriscus sylvestris Cow Parsley; Costag Fhiadhain
Absent from all treeless areas. Occasional. EAsi BoTe. Skye (14, 23-26, 32-34, 36, 44, 46, 50-55, 60-62, 72), Ra (53), Ru, E, M, C, So.

Myrrhis odorata Sweet Cicely; Mirr
Waste ground. Talisker farm; Wall, Upper Duntulm; Kyleakin. Introduced. Skye (33, 37, 72), C.

Conopodium majus Pignut; Braonan
Grassland and woods. Common. Ocea Temp. Skye (all squares), Ra (53-55, 65), Ro, Ru, E, M, C, So.

Aegopodium podagraria Ground-elder, Goutweed, Bishopweed; Lus an Easbaig
Waste ground, usually near buildings. Local. Introduced. Skye (14, 15, 24-26, 32-37, 42-47, 50-52, 60-62, 71, 72), Ra (53, 54), Ru, E, M, C, Sc, So.

Berula erecta Lesser Water-parsnip; Folachdan Beag
Marsh below bridge on pier road, Camas Mor; stream above shore, Osmigarry; burn W. of Sartle. Rare. Introduced? Euro Temp. Skye (37, 46).

Oenanthe crocata Hemlock Water-dropwort; Dàtha Bàn Iteodha
Wet places, usually near the sea. Local. SubO SoTe. Skye (all squares except 23, 26, 56), Ra (53-55, 65), Ro, Ru, E, M, C, Sc, So.

Apium inundatum Lesser Marshwort; Fualastar
Muddy edge of lochan on Canna, 1984. SubO Temp. C.

Carum verticillatum Whorled Caraway; Carbhaidh Fhàinneach
Damp runnel below road, Sligachan, 1984. Ocea SoTe. Skye (42).

Ligusticum scoticum Scots Lovage; Sunais
Around the coast, on cliff ledges, and rocky shores. Local, commoner on west side. Euro BAMo. Skye (14, 15, 23-26, 32-37, 42, 46, 47, 50, 51, 53, 54, 60, 71), Ra (53-55, 65), Ro, Ru, M, C, So.

Angelica sylvestris Wild Angelica; Lus nam Buadh
Damp grassland, woods, and even high on hills, on cliff ledges. Common. ESib BoTe. Skye (all squares), Ra (53-55, 64, 65), Ro, Ru, E, M, C, Sc, So.

Heracleum sphondylium Hogweed; Giuran
Rough grassland. Common. EAsi BoTe. Skye (all squares except 41), Ra (53, 54), Ru, E, M, C, Sc, So.

Torilis japonica Upright Hedge-parsley; Peirsill Fàil

Isolated plants in various parts of Skye often far from hedges! EAsi Temp. Skye (33, 36, 37, 51, 54, 55, 60-62, 71, 72), Ra (53, 54).

Daucus carota Wild Carrot; Curran

Grassy banks inland, and grassland near the sea. Local. ESib SoTe. Skye (all squares except 23, 32, 41, 43), Ra (53, 54), Ru, E, M, C.

Centaurium erythraea Common Centaury; Ceud-bhileach

Dry grassland near the sea, mostly in S Skye. Wall of Armadale Castle (1960s); Kilmory, Rum. Rare. Euro SoTe. Skye (36, 41, 42, 47, 50-52, 55, 60-62), Ru, E, M, C, So.

Gentianella campestris Field Gentian; Lus a'Chrùbain

Grassland, including Storr at over 2000ft; white flowered patch below Storr, 1984. Road verges. Locally common. Euro BoTe. Skye (all squares except 23, 43, 71, 72), Ra (53, 54), Ru, E, M, C, So.

G. amarella Autumn Gentian, Felwort; Lus a'Chrùbain Tuathach

Limestone grassland behind shore, Torrin (1970s); roadside bank, Suardal. Rare. Circ BoTe. Skye (51, 52, 62).

Vinca minor Lesser Periwinkle; Gille-fionn

Garden escape, Portree and Raasay. Introduced. Skye (44), Ra (53).

Solanum dulcamara Bittersweet; Searbhag Mhilis

Dunvegan Castle estate, 24. Rare. Introduced. EAsi SoTe. Skye (24, 61), Ra (53, 54), So.

Calystegia soldanella Sea Bindweed; Flùr a'Phrionnsa

Kilmory, Rum, 1991. Med Atl. Ru.

Calystegia sepium Hedge Bindweed; Dùil Mhial

Waste ground. Introduced and then naturalised. Circ Temp. Skye (14, 36, 44, 45, 47, 50-52, 60-62, 71, 72), Ra (53), Ru, E, M, C, So. Two closely related species also occur but are probably under-recorded. **C. pulchra** (*C. sepium* ssp. *pulchra)* Hairy Bindweed; Skye (44, 50, 62) and **C. silvatica** (*C. sepium* ssp. *silvatica)* Large Bindweed; Skye (24, 44, 50, 62).

Menyanthes trifoliata Bogbean; Tri-bhileach

Lochs and wet bogs. Locally common. Circ BoTe. Skye (all squares), Ra (53-55, 64, 65), Ro, Ru, E, M, C, Sc, So.

Symphytum x **uplandicum** = **S. officinale** x **S. asperum** Russian Comfrey

Garden escape. Introduced. Skye (24, 35, 36, 44, 46, 47, 51, 52, 60, 62, 72). C.

S. tuberosum Tuberous Comfrey; Meacan Dubh Cnapach
Corry Lodge, Broadford; grounds of Raasay House. Introduced.
Euro Temp. Skye (62), Ra (53).

Anchusa arvensis *Lycopsis arvensis* Bugloss; Lus Teanga an Daimh
Arable weed, Glen Brittle; Uigshader; unlocalised in 34. Rare.
ESib Temp. Skye (34, 42, 44), Ru, E, M, C.

Pentaglottis sempervirens Green Alkanet; Bog-lus
Garden escape. Dunvegan; Armadale; Dunringell, Kyleakin.
Introduced. Skye (24, 60, 72).

Mertensia maritima Oysterplant; Tiodlach na Mara
About a dozen plants on shingle at Ardroag. Known there since
the 1930s, but numbers vary - only one plant, 1978. New
colony Harlosh, 1989. Euro BAMo. Skye (24), Ru, E (until
1974), C.

Amsinckia micrantha Common Fiddleneck
Garden weed, Port Mor, Muck, 1996. Introduced. M.

Myosotis scorpioides Water Forget-me-not; Cotharach
May be confused with *M. secunda* below. Reliable records
required for both. ESib Temp. Skye (14, 23, 25, 26, 32, 34-37,
41-46, 50-52, 55, 56, 61, 62), Ra (53-55, 64), Ro, Ru, E, M, Sc,
So.

M. secunda Creeping Forget-me-not; Lus Midhe Ealaidheach
Commoner in peaty wet places and on hills. Check against
above. Ocea Temp. Skye (all squares except 41, 42, 47, 54), Ra
(53-55, 64, 65), Ro, Ru, E, M, C, Sc, So.

M. laxa *M. caespitosa* Tufted Forget-me-not; Lus Midhe Dosach
Marshes, burns and ditches. Local. Circ BoTe. Skye (14, 15,
24-26, 33-37, 42-44, 47, 50-52, 55, 61, 62), Ra (53-55), Ro,
Ru, E, M, C, Sc, So.

M. sylvatica Wood Forget-me-not; Lus Midhe Coille
Viewfield, Portree; Dunringell, Kyleakin; woods, Inverarish,
Raasay. Introduced. EAsi Temp. Skye (44, 72), Ra (53).

M. arvensis Field Forget-me-not; Lus Midhe Aitich
An uncommon weed of waste ground. ESib BoTe. Skye (14, 15,
24, 26, 32, 34-37, 44, 45, 47, 50-52, 54, 60-62, 72), Ra (53,
54), Ru, E, M, C, So.

M. discolor Changing Forget-me-not; Lus Midhe Caochlaideach
A common weed of cultivated ground. Euro Temp. Skye (all
squares except 23, 33, 41, 53), Ra (53-55, 65), Ro, Ru, E, M, C,
Sc, So.

Landscapes of North Ebudes

The Quiraing, Skye

Trallval, Askival, and
Ainshval, Rum

Sgurr nan Gillean,
Cuillin Hills, and Loch
nan Eilean, Skye

Compass Hill, Canna

(Photos John Birks)

Plants of coastal habitats

Upper left: *Asplenium marinum*, right: *Armeria maritima*;
Lower left: *Ligusticum scoticum*, right: *Scilla verna*.
(photos John Birks)

Plants of wet habitats

Upper left: *Caltha palustris*, right: *Narthecium ossifragum*;
Lower left: *Pinguicula lusitanica*, right: *Eriocaulon aquaticum*.
(photos John Birks)

Plants of mountain habitats

Upper left: *Koenigia islandica*. right: *Saussurea alpina*;
Lower left: *Saxifraga nivalis*, (photos John Birks)
right: *Arenaria norvegica*, (photo C.W. Murray)

Stachys officinalis *Betonica officinalis* Betony; Lus Beathaig
Portree (Meall), and Penefiler (Inveralivaig). Three records in
Sleat all pre-1970. Rare. Euro Temp. Skye (44, 47, 60-62).

S. sylvatica Hedge Woundwort; Lus nan Sgor
In shadier places than above. Common. ESib Temp. Skye (all
squares except 41, 53), Ra (53-55, 64, 65), Ro, Ru, E, M, C, Sc,
So.

S. x ambigua = S. sylvatica x S. palustris Hybrid Woundwort
Perhaps overlooked. Skye (44, 50, 61, 71), Ra (53), Ru, E.

S. palustris Marsh Woundwort; Brisgean nan Caorach
Damp grassland, often in arable ground. Locally common. Circ
BoTe. Skye (all squares except 23, 32, 41-43, 54), Ra (53-55),
Ru, E, M, C, Sc, So.

S. arvensis Field Woundwort; Creuchd-lus Arbhair
An uncommon arable weed. SubO SoTe. Skye (36, 45, 47, 51,
72). M, C.

Lamium purpureum Red Dead-nettle; Caoch-dheanntag Dhearg
Weed of cultivation, and in gardens in Portree. Local. Euro
Temp. Skye (24-26, 34-37, 42, 44, 47, 50, 51, 53, 60-62, 71,
72), Ra (53), Ru, E, M, C, So.

L. hybridum Cut-leaved Dead-nettle; Caoch-dheanntag Gheàrrte
Weed of cultivation, Kildonan, Eigg, 1998. E.

L. confertum Northern Dead-nettle; Caoch-dheanntag Thuathach
Weed of cultivated ground. Euro BoMo. Kilmarie 51, Broadford
62. Skye (51, 62), Ra (53), E, M, So.

Galeopsis speciosa Large-flowered Hemp-nettle; An Gath Mòr
In cultivated ground. Large flowers, yellow with purple lip. Now
rare. ESib BoTe. Skye (15, 24, 35, 44, 45, 47, 50-52, 60-62),
Ru, E, C, So.

G. tetrahit Common Hemp-nettle; Deanntag Lìn
Arable weed. Common. Euro BoTe. Skye (all squares except
23, 41, 54-56), Ra (53, 54), Ru, E, M, C, Sc, So.

G. bifida Bifid Hemp-nettle; An Gath Beag
Smaller than either of the above, and probably
under-recorded. EAsi BoTe. Skye (26, 36, 43-45, 71), Ra (53,
54), M.

Scutellaria galericulata Skullcap; Cochall
In shingle at various places round the coast. Local. ESib BoTe.
Skye (14, 23-25, 33, 34, 36, 41, 46, 47, 50-53, 55, 60-62, 71,
72), Ra (53-55, 64, 65), Ro, Ru, E, M, C, Sc, So.

S. minor Lesser Skullcap; Cochall Beag
Easily overlooked in wet heathland, especially when not in flower. Very local. SubO SoTe. Skye (23, 31-33, 36, 41, 43, 50-52, 60-62, 71), Ra (53, 55), Ro, Ru, E, M, C, Sc, So.

Teucrium scorodonia Wood Sage; Sàisde Coille
Dry banks, rock ledges, on moors and hills. Common. SubO SoTe. Skye (all squares), Ra (53-55, 64, 65), Ro, Ru, E, M, C, Sc, So.

Ajuga reptans Bugle; Glasair Choille
Woodland and scrub, burn gorges, and rock ledges on hills. Common. Euro Temp. Skye (all squares except 37), Ra (53, 54, 64, 65), Ro, Ru, E, C, Sc, So. Missing from Muck?

A. pyramidalis Pyramidal Bugle; Glasair Bheannach
Occasional on dry banks, or in rough grassland. Not recorded from Skye. Euro BoMo. Ru, E, M, C.

Glechoma hederacea Ground-ivy; Eidheann Thalmhainn
Woodland, Dunvegan Castle; Kilmarie, Strathaird; Corry, Broadford; garden weed, Portree; among other places, mostly in S Skye. Introduced? EAsi BoTe. Skye (24, 33, 36, 42, 44, 51, 60-62, 72), Ra (53), E, C.

Prunella vulgaris Selfheal; Dubhan Ceann-chòsach
Grassland and waste ground. Common. Circ WiTe. Skye (all squares), Ra (53-55, 64, 65), Ro, Ru, E, M, C, Sc, So.

Thymus polytrichus *T. drucei* Wild Thyme; Lus na Macraidh
Dry moorland, rocks, and scree. Common. Euro BoTe. Skye (all squares), Ra (53-55, 64, 65), Ro, Ru, E, M, C, Sc, So.

Lycopus europaeus Gipsywort; Feòran Curraidh
Wet stony sea shores; inland on Raasay (65). Very local. ESib Temp. Skye (24, 25, 33, 34, 50, 51, 53, 60, 61, 71, 72), Ra (53, 54, 65), Ro, E, M, C, So.

Mentha arvensis Corn Mint; Piunn
Arable fields, road verges; roadside ditch in centre of Portree over several seasons in 1970s. Now rare and disappearing. Circ BoTe. Skye (24, 33, 35, 42, 44-46, 50-52, 60-62, 71), Ra (unlocalised), Ru, E, Sc.

M. x **verticillata** = **M. arvensis** x **M. aquatica** Whorled Mint; Meannt Fàinneach
Shore at Camastianavaig, 53. Skye (24, 44, 51, 53, 62), E.

M. aquatica Water Mint; Meannt an Uisge
In burns and ditches, and on loch margins. Local. Euro Temp. Skye (all squares except 23, 34), Ra (53-55, 65), Ro, Ru, E, M, Sc.

M. spicata Spearmint; Meannt Gàrraidh
Introduced. Raasay (54) and Fladday. Ra (54, 55).

M. suaveolens *M. rotundifolia* Round-leaved Mint; Meannt Cùbhraidh
Introduced and naturalised. SubM SubA. Portree, 44 and Eigg. Skye (44), E.

Hippuris vulgaris Mare's-tail; Earball Capaill
In a pool beside the shore, Kildorais, 47; also burn E of Elgol, 51. Unconfirmed record for 62. Rare. Circ BoTe. Skye (47, 51, 62).

Callitriche hermaphroditica *C. autumnalis* Autumnal
Water-starwort; Biolair Ioc an Fhoghair
Lochs and streams. Probably under-recorded. Circ BoMo. Skye (14, 15, 26, 32, 43-47, 61), Ra (53).

C. stagnalis Common Water-starwort; Biolair Ioc
Commoner in N Skye. Euro Temp. Skye (14, 15, 23-26, 31-36, 43-47, 50-53, 55, 60-62, 71, 72), Ra (53-55, 64, 65), Ro, Ru, E, M, C, Sc, So.

C. hamulata *C. intermedia* Intermediate Water-starwort; Biolair Ioc Meadhanach
Again more records from N Skye. Possibly under-recorded in S Skye. Locally plentiful. SubO BoTe. Skye (24-26, 33-35, 43-46, 51, 55), Ra (53), Ru, E, M, C, Sc, So.

Plantago coronopus Buck's-horn Plantain; Adhairc Fèidh
On rocks and in salt-marsh and cliff-top turf. Absent where rivers enter the sea, as at head of Loch Snizort. Locally plentiful. ESib SoTe. Skye (all squares except 42, 44, 45, 54, 55), Ra (53-55, 64, 65), Ro, Ru, E, M, C, Sc, So.

P. maritima Sea Plantain; Slàn-lus na Mara
On mountain rocks as well as in salt-marshes and ground near the sea. Common. ESib WiBo. Skye (all squares), Ra (53-55, 64, 65), Ro, Ru, E, M, C, Sc, So.

P. major Greater Plantain; Cuach Phadraig
Cultivated and waste ground. Common. EAsi WiTe. Skye (all squares except 41), Ra (53-55, 65), Ro, Ru, E, M, C, Sc, So.

P. lanceolata Ribwort Plantain; Bodach Dubh
Grassy places. Common. ESib SoTe. Skye (all squares), Ra (53-55, 64, 65), Ro, Ru, E, M, C, Sc, So.

Littorella uniflora Shoreweed; Lus Bòrd an Locha
In lochs and pools with shallow margins. Locally plentiful.
SubO Temp. Skye (all squares except 23, 37, 54), Ra (53-55,
65), Ru, E, M, C, Sc, So.

Fraxinus excelsior Ash; Uinnsean
Native on the Suardal, Torrin, and Tokavaig limestone,
probably always planted elsewhere. Euro Temp. Skye (14, 15,
24, 25, 32-36, 42, 44-47, 50-53, 55, 60-62, 71, 72), Ra (53-
55), Ro, Ru, E, M, C, Sc.

Scrophularia nodosa Common Figwort; Lus nan Cnapan
Woods, grassy banks. Local, often only two or three plants.
ESib Temp. Skye (all squares except 34, 37, 47, 54), Ra (53,
54, 64, 65), Ro, Ru, E, M, C, Sc, So.

S. auriculata *S. aquatica* Water Figwort; Lus nan Cnapan Uisge
On a wall opposite the Dunvegan Hotel, 1958; still there 1993;
roadside ditch N of Roskhill, 1998. SubO SoTe. Skye (24).

Mimulus moschatus Musk; Musg
Garden escape, Duisdale 1970, Skye (71) only.

M. guttatus Monkeyflower; Meillag an Uillt
Spreads in ditches and along stream banks. Local. Introduced.
Skye (14, 15, 24, 25, 32, 34-37, 42-45, 51-53, 60-62, 71, 72),
Ra (53), E, So.

M. luteus Blood-drop-emlets; Meillag an Uillt Bhreocach
Less common in similar habitat. Balmaqueen, 47. Introduced.
Check whether **M. guttatus** x **luteus** occurs. Skye (36, 47, 56,
62).

M. variegatus x **guttatus**
Fiskavaig, 33 (Skye) only.

M x **maculosus** (*M. luteus x M. cupreus*) Scottish Monkeyflower
Differs from *M. luteus*, often with coppery-orange spots. Borve,
1997. Skye (44).

Cymbalaria muralis Ivy-leaved Toadflax; Buabh-lion Eidheanach
On estate and garden walls; Pavement edge, Portree, now gone.
Introduced. Skye (24, 35, 42, 44, 60, 71), Ra (53).

Linaria repens Pale Toadflax; Buabh-lion Liath
Garden escape, waste ground, Portree; Kyleakin. Introduced.
SubO Temp. Skye (44, 72).

Digitalis purpurea Foxglove; Meuran nan Cailleachan Marbha
Moorland, woodland, walls and rocks. Common. SubO SoTe.
Skye (all squares), Ra (53, 54, 65), Ro, Ru, E, M, C, Sc, So.

Veronica serpyllifolia Thyme-leaved Speedwell; Lus-crè Talamhainn

In grassland, and as an arable weed. Common. Circ BoTe. Skye (all squares), Ra (53-55, 64, 65), Ro, Ru, E, M, C, Sc, So. Check for spp. **humifusa** - corolla blue, inflorescence and capsule glandular - Quiraing, 47 (Skye) and Rum. Skye (47), Ru.

V. officinalis Heath Speedwell; Lus-crè Monaidh

Moorland, woods, and rocks. A large, leathery form occurs on mountains. Common. Euro BoTe. Skye (all squares), Ra (53-55, 64, 65), Ro, Ru, E, M, C, Sc, So.

V. chamaedrys Germander Speedwell; Nuallach

Grassland, woods. Common. ESib BoTe. Skye (all squares), Ra (53-55, 64, 65), Ro, Ru, E, M, C, Sc, So.

V. montana Wood Speedwell; Lus-crè Coille

Dun Liath woods, Strathaird; Kinloch woods; gorge at Brae, Raasay. Rare. Euro Temp. Skye (51, 55, 60, 61, 71), Ra (53, 54), E.

V. scutellata Marsh Speedwell; Lus-crè Lèana

Easily overlooked in marshy edges of lochs, and boggy ground. Very local. ESib BoTe. Skye (14, 15, 24-26, 33-37, 43-47, 50-52, 55, 56, 61, 62, 71), Ra (53-55, 65), Ro, Ru, E, M, C, Sc, So.

V. beccabunga Brooklime; Lochal Mothair

Most records from areas on Jurassic rocks or limestone, and nearly all from burns and marshy ground near the sea. Very local. ESib Temp. Skye (14, 25, 32, 36, 37, 42, 44, 46, 47, 51, 52, 54, 55, 61, 62), Ra (53, 54), Ro, E, M.

V. arvensis Wall Speedwell; Lus-crè Balla

Usually a weed of arable or waste ground. Local. Euro SoTe. Skye (all squares except 23, 37, 41, 54, 56), Ra (53-55, 65), Ru, E, M, C, Sc, So.

V. agrestis Green Field-speedwell; Lus-crè Arbhair

Occasional in cultivated ground. Euro Temp. Skye (42-45, 51, 60, 62), Ra (53), Ru, E, M, C, So.

V. polita Grey Field-speedwell; Lus-crè Liath

Waste ground, Kilmarie. Introduced? ESib SoTe. Skye (51, 62).

V. persica Common Field-speedwell, Buxbaum's Speedwell; Lus-crè Garraidh

Garden weed, Dunvegan Castle grounds; Lynedale; Dunringell, Kyleakin; Raasay House. Introduced. Skye (24, 35, 43, 45, 47, 50, 52, 61, 62, 72), Ra (53), E, M, C, So.

V. filiformis Slender Speedwell; Lus-crè Claidh
Escapes from estates and gardens, and spreads rapidly in grassy ground. Introduced. Skye (34-36, 43-45, 51, 60, 62, 72), Ra (53), E, M.

V. hederifolia Ivy-leaved Speedwell; Lus-crè Eidheannach
Garden weed, Dunringell, Kyleakin, 1978. Confirms unlocalised record of 1911. Two other records in Sleat. Old record from Eigg. Euro SoTe. Skye (50, 60, 72), E.

Melampyrum pratense Common Cow-wheat; Caraid Bhuidhe
Usually in hazel or birch scrub, but in wet open moorland in Glen Sligachan. Occasional. ESib BoTe. Skye (all squares except 23, 26, 32, 37, 54), Ra (53, 54), Ro, Ru, E, C, Sc, So.

Euphrasia officinalis agg. Eyebright; Lus nan Leac
Common. Skye (all squares), Ra (53-55, 64, 65), Ro, Ru, E, M, C, Sc, So. Further recording of microspecies required. The following have been found so far, and identified by P.F. Yeo or A.J. Silverside:

E. arctica ssp. **borealis** *E. brevipila*
Grassy banks, roadsides and fields. Ocea BoTe. Skye (all squares except 43), Ra (53, 54), Ru, E, M, C, Sc, So.

E. tetraquetra *E. occidentalis* *
Duntulm Castle, 47. Ocea Temp. Skye (47), Ru, E, M.

E. nemorosa
Lowland grassland. Euro Temp. Skye (23, 33, 34, 42, 45, 47, 51, 52, 61, 62, 71, 72), Ru, E, M.

E. confusa
Mountain grassland and moorland rocks. Salt-marsh at Kensaleyre, 45. Ocea BoTe. Skye (14, 24, 33, 34, 41, 42, 44-46, 51, 61, 62, 71), Ra (53), Ru, E, M, So.

E. frigida
Damp rock ledges, and burn sides on hills. Esib ArMo. Skye (41, 45, 53), Ru, E.

E. foulaensis *
Hallival, Rum only. Ocea BoMo. Ru.

E. ostenfeldii *E. curta*
Grassy mountain slopes, and near the sea. Ocea BAMo. Skye (14, 44-46, 52, 54, 55, 72), Ra(53), Ru, E.

E. marshallii
Sea coast near Ardmore Point; Duntulm. Ocea BoMo. Skye (25, 26, 47), E.

E. micrantha
>Heaths and moors. Euro Temp. Skye (all squares except 36, 37, 43, 46, 55, 60), Ra (53), Ru, E, M, Sc, So.

E. scottica
>Wet flushes on moors and hills. Euro BoMo. Skye (14, 15, 24, 25, 33-36, 41-47, 50-56, 60-62, 71, 72), Ra (53, 54), Ru, E, M.

E. heslop-harrisonii
>Salt marshes, Kensaleyre and Lower Breakish. Ocea BoMo. Skye (45, 62), Ro, Ru.

The following hybrids have been determined by A.J. Silverside:

E. arctica x **marshallii** (25, 26); **E. tetraquetra** x **E. marshallii** (47); **E. frigida** x **E. scottica** (42); **E. ostenfeldii** x **E. confusa** (45); **E. nemorosa** x **E. marshalii** (26) - all on Skye. On Rum there are five: **E arctica** x **E. nemorosa**, **E. confusa** x **E. nemorosa**, **E. nemorosa** x **E. micrantha**, **E. nemorosa** x **E. scottica**, **E. frigida** x **E. ostenfeldii**.

Odontites verna Red Bartsia; Modhalan Coitcheann
>Farm tracks and waste places. Local. EAsi Temp. Skye (all squares except 23, 32, 41-43, 54), Ra (53-55), Ru, E, M, C, Sc.

Rhinanthus minor agg. Yellow Rattle; Modhalan Buidhe
>Grassland. Common. Euro BoTe. Skye (all squares except 41), Ra (53, 54), Ru, E, M, C, Sc, So. The following subspecies occur:
>ssp. **minor** in Skye (24, 43, 46, 60, 62), Ra (53), Ru, M, So.
>ssp. **stenophyllus** in Skye (14, 15, 24, 26, 32-36, 44, 47, 52, 55, 62, 71, 72), Ra, Ru, E, M.
>ssp. **monticola** in Skye (26, 43, 56), Ru, M.
>ssp. **borealis** Rock ledges on cliffs of Storr-Quiraing ridge; Blaven. Local. Skye (45-47, 52, 55), Ra (53), Ru.

Pedicularis palustris Marsh Lousewort, Red-rattle; Lus Riabhach
>Wet ground and shallow loch edges. Common. Euro BoTe. Skye (all squares), Ra (53-55, 64, 65), Ro, Ru, E, M, C, Sc, So.

P. sylvatica Lousewort; Lus Riabhach Monaidh
>Damp moors, and boggy ground. Common. Euro Temp. Skye (all squares), Ra (53, 54, 65), Ro, Ru, E, M, C, Sc, So. Check for spp. **hibernica** - long white hairs on calyx and pedicels. Skye (14, 15, 23-25, 31-35, 37, 42-47, 50, 52, 53, 55, 56, 61), Ra (53-55, 64, 65), Ro, Ru.

Orobanche alba Thyme Broomrape; Siorralach
>Rocky slopes and sea cliffs, mainly in N Skye. Rare and often only a single plant. Euro Temp. Skye (14, 23-26, 33, 36, 44-47, 50, 51, 56, 61, 62), Ra (53, 54), Ru, E, M, C.

Pinguicula lusitanica Pale Butterwort; Mòthan Beag Bàn
Bogs and wet moorland, flowering later than *P. vulgaris*. Local. Ocea Temp. Skye (all squares except 26, 34, 37), Ra (53-55, 64, 65), Ro, Ru, E, M, C, Sc, So.

P. vulgaris Common Butterwort; Mòthan
Bogs and wet rocks. Locally common. Circ BoMo. Skye (all squares), Ra (53-55, 64, 65), Ro, Ru, E, M, C, Sc, So.

Utricularia vulgaris agg. Greater Bladderwort; Lus nam Balgan Mòr
Lochans and moorland pools. Six of these records are *U. australis* (*U. neglecta*). Occasional. ESib Temp. Skye (25, 34, 35, 42, 45-47, 52, 53, 61, 62), Ra (54, 65), Ro, Ru.

U. intermedia Intermediate Bladderwort; Lus nam Balgan Meadhanach
Shallow edges of lochs, or muddy pools. Most records in S Skye. Occasional. Circ BoMo. Skye (15, 25, 34, 42, 44, 50-52, 61, 62, 71, 72), Ra (53, 54), Ro, Ru, E, C.

U. minor Lesser Bladderwort; Lus nam Balgan Beag
In shallow peaty pools, where it sometimes flowers. Locally common. Circ BoTe. Skye (14, 24, 25, 32, 33, 35, 41-45, 47, 50-53, 56, 60-62, 71, 72), Ra (53, 54), Ru, E, M, C, Sc, So.

Campanula rapunculoides Creeping Bellfower
Rough ground, Eigg, 1998. - last seen 1930s. Introduced. E.

C. rotundifolia Harebell; Currac Cuthaige
Mysteriously rare and local - three tiny patches in Uig; one at Teangue; one in Raasay; two records still untraced. Skerinish 1993. Roadside, Lynedale, 1978. Circ BoTe. Skye (15, 35, 36, 42-45, 47, 50-52, 60-62), Ra (53), Ru, E, M, So.

Lobelia dortmanna Water Lobelia; Flùr an Lochain
Shallow edges of lochs. Common. Euro BoMo. Skye (all squares except 15, 26, 37), Ra (53-55, 64, 65), Ru, E, Sc, So.

Sherardia arvensis Field Madder; Màdar na Machrach
An uncommon weed of arable and waste ground. Rare. Casual at Prabost, 1950s (45). Euro SoTe. Skye (45, 50, 52, 54, 62), Ra (53), Ru, C.

Galium boreale Northern Bedstraw; Màdar Cruaidh
Rocky slopes and ledges; rough grassland. Locally plentiful in basalt areas, but seems to be rare in Sleat and missing from Raasay. Circ BoTe. Skye (all squares except 37, 47, 60, 71), E.

G. odoratum Woodruff; Lus a'Chaitheimh

Isolated patches in woodland, or in shade at base of basalt scarps. Local. Euro Temp. Skye (all squares except 14, 15, 37, 42, 47, 56), Ra (53, 54, 64), E, Sc, So.

G. palustre Common Marsh-bedstraw; Màdar Lèana

Marshy ground and ditches. Common. ESib BoTe. Skye (all squares), Ra (53-55, 64, 65), Ro, Ru, E, M, C, Sc, So.

G. verum Lady's Bedstraw; Lus an Leasaich

Locally common on roadside verges between Struan and Dunvegan; behind Glen Brittle beach. Isolated patches elsewhere, missing from Sleat (except 50) and Raasay. EAsi BoTe. Skye (14, 23-26, 32-36, 41-45, 47, 50-52, 56, 61, 62, 72), Ru, E, M, C, So.

G. mollugo Hedge Bedstraw; Màdar Fàil

Hayfields at Edinbane and Prabost over several seasons (1960s). Croft at Peingown, (47) 1997. Possibly Introduced. Euro BoTe. Skye (35, 45, 47).

G. saxatile Heath Bedstraw; Màdar Fraoich

Moorland, rough grassland. Common. SubO Temp. Skye (all squares), Ra (53-55, 64, 65), Ro, Ru, E, M, C, Sc, So.

G. aparine Cleavers, Sticky Willie, Goosegrass; Searcan

Often in shingle rather than in hedges. Locally common. Euro Temp. Skye (all squares), Ra (53-55, 64, 65), Ro, Ru, E, M, C, Sc, So.

Sambucus nigra Elder; Dromanach

Scattered plants in woodland, and often planted beside croft houses. Euro Temp. Skye (all squares except 23, 41, 54, 55), Ra (53, 54), Ru, E, M, C, Sc, So.

Viburnum opulus Guelder-rose; Caor-chon

Isolated specimens in rocky stream gorges - Dunvegan; Glen Brittle; Sligachan; Suardal. Locally more plentiful at Torrin, on limestone outcrops, and croft scrubland. Rare. Circ Temp. Skye (15, 24, 26, 31, 32, 34, 41-44, 51, 52, 61, 62), Ra (54), Sc.

Lonicera periclymenum Honeysuckle; Iadh-shlat

Woods, rock faces, gorges in burns. Common. SubO SoTe. Skye (all squares), Ra (53-55, 64, 65), Ro, Ru, E, M, C, Sc, So.

Valerianella locusta Common Cornsalad, Lamb's-lettuce; Leiteis an Uain

Rocks at roadside, Ord, Sleat, and in nearby garden. Rare. Euro Temp. Skye (61), Ru, E, C.

Valeriana officinalis Common Valerian; Carthan Curaidh
Damp grassland. Locally common. EAsi BoTe. Skye (all squares), Ra (53-55, 64, 65), Ro, Ru, E, C, Sc, So.

Centranthus ruber Red Valerian; Carthan Curaidh Dearg
Naturalised on Dunvegan Castle walls. Introduced. Skye (24).

Succisa pratensis Devil's-bit Scabious; Ura-bhallach
Grassland, moorland, rock ledges on hills. Common. ESib Temp. Skye (all squares), Ra (53-55, 64, 65), Ro, Ru, E, M, C, Sc, So.

Carlina vulgaris Carline Thistle; Cluaran Oir
Dry rock ledges or grassy slopes near the sea, mainly in N and W Skye. Rare. ESib Temp. Skye (14, 15, 24, 26, 33, 36, 47, 51, 61), Ru, E, M, C.

Arctium minus agg. Lesser Burdock; Cliadan
Waste places. Occasional. EAsi Temp. Skye (14, 15, 24, 26, 32, 33, 35, 41, 43, 47, 50, 51, 53, 54, 60, 61, 71), Ra (53-55, 64, 65), Ro, Ru, E, M, C, So. This taxon has recently been split into two species, *A. minus* and *A. nemorosum* by Stace (1997). It is likely that our records are mainly *A. minus* s.s. but more work is needed.

Saussurea alpina Alpine Saw-wort; Sàbh-lus Ailpeach
Scattered patches, or even single plants, on mountain ledges, or in scree. Storr-Quiraing ridge; Cuillin; Blaven group; Kyleakin hills. Also on cliffs above sea at Talisker. Raasay, 2003. Rare except in Cuillin. EAsi ArMo. Skye (15, 24, 32, 41, 42, 45-47, 52, 53, 71, 72), Ra (53), Ru.

Cirsium vulgare Spear Thistle; Cluaran
Fields, waste places. Common. ESib Temp. Skye (all squares), Ra (53-55, 64, 65), Ro, Ru, E, M, C, Sc, So.

C. heterophyllum Melancholy Thistle; Cluas an Fheidh
Damp grassy banks by burns; damp fields. Local. ESib BoMo. Skye (all squares), Ra (53, 54, 64), E.

C. palustre Marsh Thistle; Cluaran Lèana
Wet grassland, ditches. Common. ESib BoTe. Skye (all squares), Ra (53-55, 64, 65), Ro, Ru, E, M, C, Sc, So.

C. arvense Creeping Thistle; Fothannan Achaidh
Fields, waste places. Common. EAsi Temp. Skye (all squares except 23, 41), Ra (53-55), Ro, Ru, E, M, C, Sc, So.

Centaurea nigra Common Knapweed; Cnapan Dubh
Rough grassland. Common. SubO Temp. Skye (all squares), Ra (53-55, 64, 65), Ro, Ru, E, M, C, Sc, So.

Lapsana communis Nipplewort; Duilleag Bhràghad
Edges of woodland, and also an arable weed. Occasional. Euro Temp. Skye (14, 24-26, 33, 35, 36, 43-45, 47, 50-54, 56, 60-62, 71, 72), Ra (53, 54), Ru, E, M, C, So.

Hypochoeris radicata Cat's-ear; Cluas Cait
Grassland; moorland and roadside banks. Common. Euro SoTe. Skye (all squares), Ra (53-55, 64, 65), Ro, Ru, E, M, C, Sc, So.

Leontodon autumnalis Autumn Hawkbit; Caisearbhan Coitcheann
Grassland, river shingle, and screes. Common. Euro BoTe. Skye (all squares except 41), Ra (53-55, 64, 65), Ro, Ru, E, M, C, Sc, So.

L. saxatilis *L. taraxacoides* Lesser Hawkbit; Caisearbhan as Lugha
Roadside verge at Storr Lochs Dam in 1960s, 55; even older unconfirmed records in 41, 50. Rare. SubO SoTe. Probably introduced. Skye (41, 50, 55), Ru, E, C (also old records).

Sonchus arvensis Perennial Sow-thistle; Giogan Tolltach
An arable weed, or on marshy shores. Missing from SE Skye. Local. ESib Temp. Skye (24, 25, 35, 36, 41, 42, 44, 45, 47, 50, 60), Ra (53), Ru, E, C.

S. oleraceus Smooth Sow-thistle; Bainne Muice
Large gaps in distribution, possibly a result of confusion with the other two species. Euro SoTe. Skye (23-25, 33, 34, 44, 45, 50, 51, 54, 60-62, 72), Ra (53, 54), Ru, E, M, C, Sc, So.

S. asper Prickly Sow-thistle; Searbhan Muice
Arable weed. Locally common. Euro SoTe. Skye (14, 15, 23-26, 33, 34, 36, 37, 42, 43, 45, 50-52, 55, 56, 60-62, 71, 72), Ra (53-55, 64, 65), Ro, Ru, E, M, C, Sc, So.

Cicerbita macrophylla ssp. **uralensis** Common Blue Sow-thistle
Garden escape in ditch at Camas Mor, Bornaskitaig, 1960s, now gone. Introduced. Skye (37).

Mycelis muralis Wall Lettuce; Bliotsan
Walls of ruined outbuildings, Corry Lodge, Broadford, and on walls along 'Hospital' road nearby; Kyleakin. Introduced? Euro Temp. Skye (62, 72).

Taraxacum officinalis agg. Dandelion; Beàrnan Brìde
The aggregate is locally common everywhere. Circ WiTe. Division into the following sections is required:
Section **Erythrosperma**: Dry places, sandy heaths, dunes.
Taraxacum brachyglossum Skye (42).
T. haworthianum Skye (33).

T. fulvum Skye (35).
>Section **Obliqua**: Sand dunes and turf near the sea.
>Section **Palustria**: Damp grassland
>Section **Spectabilia**: Grassland and rocky areas

T. faeroense Skye (25, 32, 33, 35, 36, 41-45, 50, 52, 54, 55, 60, 62, 72), Ra(54), Ru, E, M.
>Section **Naevosa**: Grassland and rocky areas

T. naevosum Skye (42, 45, 72).

T. naevosiforme Skye (42, 72).

T. euryphyllum Skye (36, 42, 45, 62).

T. maculosum Skye (25, 37, 44, 45, 52, 54, 60-62), Ra (54), E, M.

T. subnaevosum Skye (51, 52), E, M.

T. drucei Skye (46, 51, 52), E.

T. stictophyllum Skye (47), E.
>Section **Taraxacum**: Basic mountain rock ledges
>Section **Celtica**: Damp grassland and rock ledges

T. gelertii Skye (45).

T. subbracteatum Skye (62).

T. duplidentifrons Skye (24, 36, 42, 44), Ra (53).

T. landmarkii Skye (25, 36, 42, 44, 45, 50-52, 61, 62, 72), E.

T. ostenfeldii Skye (72).

T. fulvicarpum Skye (25). Ru.

T. unguilobum Skye (24, 25, 32, 33, 42, 44, 50-52, 60-62, 72), Ra (53), E, M.
>Section **Hamata**: Damp grassland and roadsides

T. hamatum Skye (44).

T. lamprophyllum Skye (33).
>Section **Ruderalia**: Grassland and waste ground

T. insigne Skye (24).

T. sellandii Skye (44, 52).

T. cordatum Skye (44).

T. ekmanii Skye (44).

T. cophocentrum Skye (44).

T. dahlstedii Skye (62).

T. polyodon Skye (44, 45, 60), E.

T. xanthostigma Skye (72).

All records det. A. J. Richards, C. C. Haworth, or A. Dudman.

Crepis paludosa Marsh Hawk's-beard; Lus Curain Lèana
>Wet grassland, wet woods, and rocks by burns. Local. Euro BoTe. Skye (26, 32-37, 43-46, 50-53, 55, 56, 60-62, 71, 72), Ra (53, 54, 65), Ro, Ru, M, Sc.

C. capillaris Smooth Hawk's-beard; Lus Curain Mìn
Grassland, waste places. Locally common. Euro Temp. Skye (14, 15, 24, 26, 33-36, 42-47, 50-53, 56, 60-62, 71, 72), Ra (53, 54), Ro, Ru, E, M, C, Sc, So.

Pilosella officinarum *Hieracium pilosella* Mouse-ear Hawkweed; Srubhan na Muice
Dry grassy and rocky banks. Euro Temp. Skye (all squares), Ra (53, 54, 64), Ro, Ru, E, M, C, Sc, So.

P. aurantiaca ssp. **carpathicola** *H. brunneocroceum, H. aurantiacum* ssp. *carpathicola*
Naturalised in Dunvegan, Portree and Kyleakin. Skye (24, 44, 62, 72), Ra (53).

Hieracium vulgare agg. Hawkweed; Lus na Seabhaig
Grassy banks, edges of burns, rocky gorges, sea cliffs, and mountain rocks and ledges. The following 47 species have been recorded so far, all of which were identified or confirmed by P. D. Sell and C. West; later by J. Bevan or D. McCosh.

Hieracium holosericeum
Rocks and scree on hills. Lightfoot's record of 1772 from Beinne na Greine refound 1986. Also Ben Aslak, 1991. Skye (71, 72).

H. lingulatum
Blaven, 52. Kyleakin hills, 72. Skye (52, 72).

H. vennicontium *
Storr cliffs, Skye (45).

H. dasythrix
Kyleakin hills, Skye (72).

H. petrocharis
Rocky outcrop, Tote (1960s). Skye (45).

H. shoolbredii
Rocky ground near the sea; also on limestone at Suardal. The most widespread species on Skye. Skye (14, 24, 25, 32, 33, 35, 36, 41-43, 45-47, 50-52, 54-56, 60-62, 71), Ra (53, 54), Ru, E, M, Sc.

H. hebridense
Rocky outcrops. Skye (36, 42, 45, 50, 55, 61, 62), Ra (53, 54). Ru.

H. ampliatum
Rocks, ranging from seashore to mountain cliff ledges. Skye (14, 25, 45, 47, 51, 52, 61, 62), Ra (54).

H. langwellense
Rocks and scree on hills; shore rocks at Oskaig, Raasay. Skye (44, 45-47, 54, 61), Ra (53).

H. anglicum
Grassy banks and rock ledges. Skye (15, 24, 36, 42-46, 51, 52, 54, 55), Ra (53, 54), Ru, E, C.

H. iricum
Sea cliffs, and walls. Skye (33, 36, 42, 46, 52, 55, 62), E, C, Sc.

H. stenopholidium *
Limestone at Hallaig, Raasay. Ra (53).

H. ebudicum
Most from shore cliffs, all limestone or Jurassic rocks. Skye (25, 36, 41, 46, 61), Ra (53).

H. subglobosum
Rocks and scree on hills. Beinne na Greine 1989. Ben Aslak 1991. Probably Lightfoot's *H. alpinum* of 1772. Skye (71, 72).

H. schmidtii
Bank of burn, head of Loch Ainort; rocks above Quiraing road; Sanday, Canna. Skye (46, 52), C.

H. nitidum
Cliffs above shores, or rocky stream gorges; Brochel Castle, Raasay. Skye (36, 43, 46, 60, 71), Ra (54).

H. jovimontis *
Grassy bank by the sea. Skye (53).

H. argenteum
Stream gorges, or inland cliffs. Skye (33, 42, 44, 50-53, 61, 62, 72), Ra (53, 54), Ru.

H. caledonicum
Rocks by rivers or streams. Skye (14, 24, 25, 34, 36, 41-43, 45, 46, 51, 52, 55, 61, 72), Ra (53, 54). Ru, E, Sc.

H. subrude
Inland cliff, Uig; coastal cliff, Knock, Sleat. Skye (36, 60).

H. orimeles
Cliffs and rocks by streams. Skye (14, 25, 42, 44, 53, 72).

H. chloranthum
Rocks by rivers and shores. Skye (24, 26, 32, 36, 42-46, 51, 72), Ra (53), E.

H. exotericum
Roadside rocks, Loch Cill Chriosd 1970. Skye (62).

H. uistense
Shore rocks and inland cliffs on hills. Skye (42, 44, 45, 51, 52, 55, 61, 62), Ru.

H. duriceps
Grassy banks, rock ledges. Skye (24, 25, 41-43, 45, 51, 52, 61, 62, 72), Ra (54), E.

H. pictorum
Rocky stream banks, shore cliffs. Skye (24, 42, 51, 52), Ra (53, 54).

H. pollinarioides *
Limestone cliffs and rock ledges, Raasay. Ra (53, 54).

H. piligerum *
Rock ledges. Near Elgol, 51; Blaven 52. Skye (51, 52).

H. cymbifolium
Shore cliffs, Skye; limestone cliffs and rock ledges, Raasay. Skye (51), Ra (53, 54).

H. pseudostenstroemii
Hallaig, Raasay; Rudh'a'Chinn Mhoir, Scalpay. Ra (53), E, Sc.

H. subtenue
Cliff, Sgurr Mor. Skye (47).

H. subhirtum
Cliffs and rocky stream-sides. Skye (36, 45), Ra (54).

H. rhomboides *
Suardal. Skye (62).

H. dipteroides
Limestone bank, Torrin. Skye (52).

H. caesiomurorum *
Limestone cliffs, Raasay; Fionchra, Rum. Ra (53), Ru.

H. orcadense *H. euprepes*
Hill and mountain rock ledges. Skye (45-47, 51, 55, 62), E.

H. diaphanoides *
Limestone gorge, Druim an Aonach, Raasay. Ra (54).

H. rubiginosum
Coastal cliffs, and inland by burns. Skye (14, 24, 25, 33, 36, 42, 47, 61, 72), Ra (54).

H. vulgatum
Rocky and grassy places, commonest after *H. shoolbredii.* Skye (33, 36, 41-45, 50-52, 55, 56, 60-62, 71, 72), Ra (53, 54), E, M, So.

H. cravoniense
Cliffs and rocky places, most so far on limestone. Skye (36, 51-53, 62).

H. sparsifolium
Rock ledges, roadsides, river banks. Skye (15, 24, 36, 44, 50, 54).

H. uiginskyense
Rocks and roadside banks. Skye (25, 35, 36, 44).

H. latobrigorum
Cliffs, walls, fields. Skye (33, 36, 46, 47, 51-53, 61), Ra (53), Sc.

H. subcrocatum
Cliffs, grassy banks. Skye (25, 36, 44-46, 60), E, M.

H. strictiforme
Seashore and inland cliffs, grassy banks. Skye (26, 33, 35, 36, 46, 47, 50, 60), Ra (54), E, M, Sc.

H. reticulatum
Eynort, 32, Tokavaig ravine, 61. Skye (32, 61).

H. subaudum
Woodland, Kilbride House. Skye (52). Gone by 1989.

Antennaria dioica Mountain Everlasting, Cat's-foot; Spog Cait
On rocks and thin soil on moorland. Check whether var. *hyperborea* occurs. Local. EAsi BoTe. Skye (all squares), Ra (53-55, 64, 65), Ro, Ru, E, M, C, Sc, So.

Anaphalis margaritacea Pearly Everlasting
Garden escape. Introduced. Skye (24, 35, 45).

Gnaphalium sylvaticum Heath Cudweed; Cnàmh-lus Mòintich
On dry or stony paths or in rough grass. Rare. ESib BoTe. Skye (15, 24, 35, 45, 50-53, 60-62, 71, 72), Ra (53-55, 64), Ru, So.

G. supinum Dwarf Cudweed; Cnàmh-lus Beag
In scree or stony ground high on Storr-Quiraing ridge; Cuillin; Red Hills; Blaven; Sgurr na Coinnich. Local to rare. Euro ArMo. Skye (42, 45-47, 51-53, 62, 71, 72).

G. uliginosum Marsh Cudweed; Cnàmh-lus Lèana
Isolated patches on damp paths, dried-up pond edges, etc. Rare. EAsi BoTe. Skye (24, 25, 33-35, 44, 45, 47, 50, 51, 60-62, 71, 72), Ra (53, 54), Ru, E, M, C.

Inula helenium Elecampane; Aillean
Dunvegan Castle grounds; Portree; Brochel, Raasay. Rare. Introduced. Skye (24, 44), Ra (54), C, Sc.

Pulicaria dysenterica Common Fleabane; Fuath Dheargann
Garden escape, ditch near Ostaig, Sleat. ESib SoTe. Skye (60).

Solidago virgaurea Goldenrod; Slat Oir

Moorland and woodland banks, mountain rock ledges. Common. EAsi BoTe. Skye (all squares), Ra (53-55, 64, 65), Ro, Ru, E, M, C, Sc, So.

Aster tripolium Sea Aster; Neòinean Sàilein

Locally plentiful in salt-marshes, but absent where sizeable rivers enter the sea (head of Loch Snizort, head of Loch Slapin). EAsi Temp. Skye (24, 25, 33-35, 44, 51, 60-62, 71, 72), Ra (53).

Bellis perennis Daisy; Neòinean

Common in grassland and as a garden weed. Euro Temp. Skye (all squares), Ra (53-55, 64, 65), Ro, Ru, E, M, C, Sc, So.

Tanacetum parthenium Feverfew; Meadh Duach

Run wild in old Schoolhouse garden, Kyleakin, 72; also 36 (Skye). Introduced. Skye (36, 72), Ra (53).

T. vulgare *Chrysanthemum vulgare* Tansy; Lus na Frainge

Probably introduced, since usually found near old houses. Occasional. EAsi BoTe. Skye (14, 15, 24-26, 33, 35-37, 44-47, 50, 52, 53, 60-62, 71, 72), Ra (53), Ru, E, M.

Artemisia vulgaris Mugwort; Liath-lus

Waste places, road verges. Local and never in any quantity. ESib Temp. Skye (14, 24, 35-37, 44, 45, 47, 50-53, 60, 62, 72), E, C.

Achillea ptarmica Sneezewort; Cruaidh-lus

Damp grassland and marshes. Common. EAsi BoTe. Skye (all squares), Ra (53-55, 64, 65), Ro, Ru, E, M, C, Sc, So.

A. millefolium Yarrow; Eàrr-thalamhainn

Drier grassland, road verges, grassy banks. Common. EAsi BoTe. Skye (all squares), Ra (53-55, 65), Ro, Ru, E, M, C, Sc, So.

Chrysanthemum segetum Corn Marigold; Buidheag an Arbhair

Weed of cultivated ground, indicating lack of lime. Formerly locally common. Introduced. Euro SoTe. Skye (14, 15, 24-26, 33-37, 42, 44-47, 50-53, 56, 60-62), Ra (54), Ru, E, M, Sc.

Leucanthemum vulgare *Chrysanthemum leucanthemum* Ox-eye Daisy; Neòinean Mòr

Hayfields and grassy banks. Common. ESib BoTe. Skye (all squares except 23, 54, 55), Ra (53, 54), Ru, E, M, C, Sc, So.

Matricaria discoidea Pineappleweed; Lus Anainn

Frequent in trampled ground round farm buildings and gateways, and on paths. Introduced. Skye (all squares except 23, 41, 54), Ra (53, 54), Ru, E, M, C, Sc, So.

Tripleurospermum maritimum Sea Mayweed; Buidheag na Mara
Shingle beaches and maritime grassland. Local. Check on occurrence of **T. inodorum** as distinct from **T. maritimum**. Circ WiBo. Skye (all squares except 43, 54), Ra (53, 54, 65), Ro, Ru, E, M, C, Sc, So.

Senecio jacobaea Common Ragwort; Buaghallan
Rough grassland, waste ground. Locally common. ESib Temp. Skye (all squares), Ra (53-55, 64, 65), Ro, Ru, E, M, C, Sc, So.

S. aquaticus Marsh Ragwort; Caoibhreachan
Marshy ground and ditches. Common. Euro Temp. Skye (all squares except 41, 54), Ra (53-55), Ro, Ru, E, M, C, Sc, So.

S. vulgaris Groundsel; Grunnasg
In cultivated and waste ground. Locally plentiful - too plentiful! Euro SoTe. Skye (all squares except 23, 25, 41, 46, 54), Ra (53, 55), Ru, E, M, C, Sc, So.

S. sylvaticus Heath Groundsel; Grunnasg Monaidh
Quarry near Ullinish; wall of ruined cottage, Gleann Meadhonach; also 44, 61. Rare. Euro Temp. Skye (33, 44, 60, 61, 62), Ra (53).

Tussilago farfara Colt's-foot; Cluas Liath
Among stones at edges of rivers and burns, or on damp muddy banks. Local. ESib BoTe. Skye (all squares except 23, 41), Ra (53, 54), Ru, E, M, Sc.

Petasites hybridus Butterbur; Gallan Mòr
Banks of burns and other wet places. Local. Euro Temp. Under-recorded or missing from S Skye. Skye (14, 23-25, 32-35, 42-47, 53, 55, 56, 60, 62), Ra (54), E, M, So.

Eupatorium cannabinum Hemp-agrimony; Cainb-uisge
Damp places on or below sea cliffs, more often on Jurassic rocks than basalt. Locally plentiful, as at Talisker. Euro Temp. Skye (24, 25, 33, 35, 36, 41, 50, 51, 53-56, 60, 61, 71), Ra (53-55, 64, 65), Ro, E, M, C, Sc.

Baldellia ranunculoides Lesser Water-plantain; Corr-chopag Bheag
Loch margins, usually submerged. Lochan Coir'a'Ghobhainn; Loch a'Ghlinne; Loch Cill Chriosd; Loch Airidh. Rare. SubO SoTe. Skye (41, 50, 60, 62).

Alisma lanceolatum Narrow-leaved Water-plantain; Corr-chopag
Introduced to man-made lochan, Lynedale estate, c. 1900. The dam burst in 1978, and although loch is gone, plants survived in the original burn. Not found elsewhere, though there are other 'constructed' fishing lochs in Skye. ESib SoTe. Skye (35).

Triglochin palustre Marsh Arrowgrass; Bàrr a'Mhilltich Lèana
Marshy ground, and edges of burns. Common. Circ BoTe. Skye (all squares), Ra (53-55, 64, 65), Ro, Ru, E, M, C, Sc, So.

T. maritimum Sea Arrowgrass; Bàrr a'Mhilltich Mara
Salt marshes, and wet places on rocky shores. Common. Circ BoTe. Skye (all squares except 23, 54-56), Ra (53, 54), Ro, Ru, E, M, C, Sc, So.

Potamogeton natans Broad-leaved Pondweed; Duileasg na h-Aibhne
Lochs and pools. Common. Circ BoTe. Skye (all squares except 15, 36, 37, 41, 55, 71), Ra (53-55), Ro, Ru, E, M, C, So.

P. polygonifolius Bog Pondweed; Liobhag Bogaich
Shallow bog-pools and ditches. Common. SubO Temp. Skye (all squares), Ra (53-55, 64, 65), Ro, Ru, E, M, C, Sc, So.

P. coloratus Fen Pondweed; Liobhag Rèisg
Loch Cill Chriosd (calcareous water). Euro SoTe. Skye (62).

P. lucens Shining Pondweed; Liobhag Loinnreach
Loch Cill Chriosd. Rare. ESib Temp. Skye (62).

P. gramineus Various-leaved Pondweed; Liobhag Fheurach
Lochs, including Loch Cill Chroisd on limestone. Local. Circ BoTe. Skye (14, 25, 44-47, 50, 51, 56, 60-62), Ra (54), Ru, E.

P. x nitens = P. gramineus x P. perfoliatus Bright-leaved Pondweed
Lochs; also pool in bend of river at Skeabost, 44. Rare. Skye (14, 44, 45, 47, 51, 56, 61, 62), Ru.

P. alpinus Red Pondweed; Liobhag Dhearg
Lochs, including Loch Cill Chriosd. Occasional. Circ BoMo. Skye (14, 25, 26, 34, 44-47, 50, 51, 55, 62).

P. praelongus Long-stalked Pondweed; Liobhag Fhada
Lochs. Rare. Circ BoMo. Skye (14, 26, 44, 45, 56, 60, 61, 71), Ra (53, 54), Ru.

P. perfoliatus Perfoliate Pondweed; Dreimire Uisge
Lochs. Commoner in NW Skye, possibly due to better recording. Circ BoTe. Skye (14, 24-26, 33-35, 44-47, 51, 53, 56), Ra (53, 54), Ru.

P. pusillus Lesser Pondweed; Liobhag Mhion
Loch Papadil, Rum, only; old record needs confirmation. Circ SoTe. Ru.

P. berchtoldii Small Pondweed; Liobhag Bheag
Loch Mor, Waterstein 14; Loch Cuithir (diatomite loch) 45. Overlooked elsewhere? Circ BoTe. Skye (14, 25, 34, 45, 47), Ra (53).

P. crispus Curled Pondweed; Lìobhag Chamagach
Loch Dubhar-sgoth. Rare. EAsi SoTe. Skye (46).

P. filiformis Slender-leaved Pondweed; Lìobhag Chaol
Loch Mor, Waterstein; Loch a'Chadhacharnaich and Loch na
Meilich, Raasay. Rare. Circ BoMo. Skye (14), Ra (53).

Groenlandia densa Opposite-leaved Pondweed; Linne-lus dluth
Introduced to man-made ponds, Skeabost House Hotel. Euro
Temp. Skye (44).

Ruppia maritima Beaked Tasselweed; Snàth-lus Mara
Brackish pools in salt-marshes. Plentiful at Camas a'Mhoir
Bheoil, Braes. Circ WiTe. Skye (31, 33, 43-45, 53), Ra (53), Ru,
M, C.

R. cirrhosa *R. spiralis* Spiral Tasselweed; Snàth-lus Camagach
Loch na h'Airde (brackish) on Rubh'an Dunain, 31. Rare. Circ
WiTe. Skye (31).

Zostera marina Eelgrass; Bilearach
Found washed up on sandy shores. Talisker; Balmeanach,
Braes; Glen Brittle; etc. Occasional. Circ WiTe. Skye (23, 25,
26, 33, 42, 50-53, 61, 62), Ra (53, 54, 65), Ru, E, M, C, Sc, So.

Lemna minor Common Duckweed; Mac gun Athair
In shallow pools and ditches. Very local; four of the seven Skye
records are from the same area. Circ SoTe. Skye (14, 25, 36,
37, 46, 47, 60), Ru, E, M, C, So.

Eriocaulon aquaticum *E. septangulare* Pipewort; Pioban Uisge
In shallow water in dozens of small lochs in the Sligachan
area, extending down Glen Sligachan as far as pools in 4924;
also in Loch Airidh na Saorach group, beside Broadford -
Armadale road; Loch nam Madadh Uisge, Luib; Loch
Scamadal, Trotternish (Butterfield & Bell, 1996). Locally
plentiful. Ocea BoMo. Skye (42, 43, 52, 55, 62).

Juncus squarrosus Heath Rush; Luachair Riasg
Damp moorland. Common. SubO Temp. Skye (all squares), Ra
(53-55, 64, 65), Ru, E, M, C, Sc, So.

J. tenuis Slender Rush; Luachair Chaol
Roadsides and waste ground. Commoner in S Skye - or
possibly under-recorded in N. Local. Introduced. Skye (24, 35,
36, 42-45, 50-52, 60-62, 72), Ra (53, 54), Ru, M.

J. gerardii Saltmarsh Rush; Luachair Rèisg Ghoirt
Salt-marshes. Locally common. Circ WiTe. Skye (all squares
except 54-56), Ra (53-55, 64, 65), Ro, Ru, E, M, C, Sc, So.

J. trifidus Three-leaved Rush; Luachair Thrì-bhileach

On rock ledges and on stony ground high on Cuillin hills; Blaven group; Glamaig; Sgurr na Coinnich; etc. Very rare compared with Cairngorms. ESib ArMo. Skye (42, 45, 52, 53, 55, 62, 71, 72), Ru.

J. bufonius agg. Toad Rush; Buabh-luachair

Muddy paths, fields and loch margins. Common. Circ WiTe. Skye (all squares except 41, 54), Ra (53-55, 65), Ro, Ru, E, M, C, Sc, So.

J. articulatus Jointed Rush; Lachan nan Damh

Wet places. Common. ESib SoTe. Skye (all squares), Ra (53-55, 64, 65), Ro, Ru, E, M, C, Sc, So.

J. acutiflorus Sharp-flowered Rush; Luachair a'Bhlath Ghèir

Wet fields and moors, on more acid soils. Euro Temp. Skye (all squares except 23, 46, 47), Ra (53, 54), Ro, Ru, E, M, C, Sc, So.

J. bulbosus Bulbous Rush; Luachair Bhalgach

Wet moorland, muddy tracks. Common. Euro BoTe. Skye (all squares), Ra (53-55, 64, 65), Ro, Ru, E, M, C, Sc, So.

J. biglumis Two-flowered Rush; Luachair Dà-lus

Wet stony places on Storr-Quiraing ridge. Check that specimen is *not* a two-flowered *J. triglumis* - lowest bract should exceed inflorescence. Rare. Circ ArMo. Skye (45-47), Ru.

J. triglumis Three-flowered Rush; Luachair Thrì-lusan

Similar habitat to above, but more plentiful. Cuillin (Fionn Corrie); Storr-Quiraing ridge. Local. Circ ArMo. Skye (42, 45-47, 55), Ru.

J. effusus Soft Rush; Luachair Bhog

Wet fields, bogs and ditches. Common. Euro SoTe. Skye (all squares), Ra (53-55, 64, 65), Ro, Ru, E, M, C, Sc, So.

J. effusus var. **spiralis**

Loch Mor, Waterstein. Probably under-recorded as it is not recognized by Stace (1977). Skye (14), Ra (54, 55, 64, 65), Ro.

J. conglomeratus Compact Rush; Luachair Dhluth

On more acid soil than *J. effusus*, and easily confused with it. Probably as common. Euro Temp. Skye (all squares except 36, 53), Ra (53-55, 64, 65), Ro, Ru, E, M, C, Sc, So.

Luzula pilosa Hairy Wood-rush; Learman Fionnach

Woods and grassy banks - more obvious in May-June than later on. Occasional. ESib BoTe. Skye (all squares except 37, 42, 52), Ra (53, 54, 64, 65), Ro, Ru, E, M, Sc, So.

L. sylvatica Great Wood-rush; Luachair Coille
More often on hill slopes and on sea cliffs than in woods. Locally plentiful. Euro Temp. Skye (all squares), Ra (53-55, 64, 65), Ro, Ru, E, M, C, Sc, So.

L. spicata Spiked Wood-rush; Learman Ailpeach
Scree and rock ledges on hills. Cuillin; Storr-Quiraing ridge; Red Hills; two hills in Sleat. Rare. Euro ArMo. Skye (42-46, 52, 54, 55, 62, 71, 72).

L. campestris Field Wood-rush; Learman Raoin
One of the first plants to flower, on turf dykes and dry grassy slopes. Locally common. Euro Temp. Skye (all squares), Ra (53-55, 64, 65), Ro, Ru, E, M, C, Sc, So.

L. multiflora Heath Wood-rush; Learman Monaidh
Rough grassland and woods. Locally common. Circ WiBo. Skye (all squares), Ra (53-55, 64, 65), Ro, Ru, E, M, C, Sc, So.

Eriophorum angustifolium Common Cottongrass; Canach
Wet moorland. Common. Circ WiBo. Skye (all squares), Ra (53-55, 64, 65), Ro, Ru, E, M, C, Sc, So.

E. latifolium Broad-leaved Cottongrass; Canach an-t-Slèibh
Wet moorland, bogs and fens, less acid than above. Rare. Euro BoTe. Skye (36, 41-45, 50-53, 55, 56, 61, 62, 71, 72), Ra (53, 54, 64), Ru, E, Sc.

E. vaginatum Hare's-tail Cottongrass; Sìoda Monaidh
Wet moorland. Less common than *E. angustifolium.* Circ BAMo. Skye (all squares), Ra (53-55, 64, 65), Ro, Ru, E, M, C, Sc, So.

Trichophorum cespitosum ssp. **germanicum** (*Scirpus cespitosus* ssp. *germanicus*) Deergrass; Ciob
Wet peaty moorland. Very common. Circ BoMo. Skye (all squares), Ra (53-55, 64, 65), Ro, Ru, E, M, C, Sc, So.

T. cespitosum nothossp. **Foersteri** *
This recently described taxon is thought to be a hybrid between ssp. *germanicum* and ssp. *cespitosum.* Although the rare ssp. *cespitosum* is not in the Hebrides, the hybrid occurs on Skye, Canna, and Raasay. Skye (51), Ra (53, 65), C.

Eleocharis palustris Common Spike-rush; Bioran Coitcheann
Marshes, ditches, and edges of lochs. Common. EAsi WiTe. Skye (all squares), Ra (53-55, 64, 65), Ro, Ru, E, M, C, Sc, So.

E. uniglumis Slender Spike-rush; Bioran Caol
Bogs, salt-marshes. Easily confused with small *E. palustris* and *E. multicaulis*; still under-recorded. Circ Temp. Skye (31, 32, 34, 41, 43, 52, 53, 55, 60-62, 71), Ra (53), Ru, E, C.

E. multicaulis Many-stalked Spike-rush; Bioran Badanach
Wet peaty places. Less common - or under-recorded? SubO Temp. Skye (14, 15, 23-25, 32, 33, 35, 37, 41, 44, 47, 50-53, 55, 60-62, 71, 72), Ra (53-55, 64, 65), Ro, Ru, E, M, C, Sc, So.

E. quinqueflora Few-flowered Spike-rush; Bioran nan Lusan Gann
Boggy moorland. Locally common. Euro BoTe. Skye (all squares), Ra (53-55, 64, 65), Ro, Ru, E, M, C, Sc, So.

Bolboschoenus maritimus *Scirpus maritimus* Sea Club-rush; Bròbh
Edges of salt-marshes. Rare, N and W Skye; commoner in Sleat. ESib SoTe. Skye (15, 24, 25, 36, 41, 43, 44, 50, 60, 61, 71), E, M, C.

Schoenoplectus lacustris *Scirpus lacustris* Common Club-rush; Luachair Ghòbhlach
Plentiful in some lochs. absent from very many more. Local. ESib WiTe. Skye (24, 34, 41, 43, 44, 46, 47, 50, 51, 53, 56, 60-62), Ra (54), Sc.

S. tabernaemontani Grey Club-rush; Luachair Bhogain
Loch na h'Airde (brackish), on Rubh'an Dunain, 31 (Skye), and on Muck. EAsi SoTe. Skye (31), M.

Isolepis setacea *Scirpus setaceus* Bristle Club-rush; Curcais Chalgach
Muddy paths and tracks. Fairly common. ESib Temp. Skye (all squares), Ra (53-55), Ro, Ru, E, M, C, Sc, So.

I. cernua *Scirpus cernus* Slender Club-rush; Curcais Chaol
Several sites on Muck, 2000. Med Atl. M.

Eleogiton fluitans *Scirpus fluitans* Floating Club-rush; Curcais air Bhog
In acid lochs and pools, and in burns. Local. Possibly under-recorded in Sleat and NE Skye. Ocea SoTe. Skye (14, 23-26, 33-35, 41-46, 50-53, 60-62), Ra (53-55, 64, 65), Ro, Ru, E, M, C, Sc, So.

Blysmus rufus Saltmarsh Flat-sedge; Seisg Rèisg Ghoirt
Salt-marshes. Locally common. Euro BoMo. Skye (14, 15, 24, 25, 31-37, 43-47, 50-53, 60-62, 71, 72), Ra (53), Ro, Ru, E, M, C, Sc.

Schoenus nigricans Black Bog-rush; Seimhean Dubh
Base-rich flushes and bogs; often on wet basic sea cliffs. Locally plentiful. ESib SoTe. Skye (all squares), Ra (53-55, 64, 65), Ro, Ru, E, M, C, Sc, So.

Rhynchospora alba White Beak-sedge; Gob-sheisg
Boggy moorland. Difficult to identify when not in flower. Commoner in more acid areas. Circ BoTe. Skye (14, 15, 24, 25, 32-36, 41-43, 45, 46, 50-53, 55, 56, 60-62, 71, 72), Ra (53-55, 64, 65), Ro, Ru, Sc, So.

R. fusca Brown Beak-sedge; Gob-sheisg Ruadh
Wet boggy ground, but very rare. SubO BoTe. Glen Sligachan 1995; Kinloch Glen, Rum. Skye (42), Ru.

Cladium mariscus Great Fen-sedge; Sàbh-sheisg
Pool off Brochel road, Raasay; also Rona and Soay. Rare. ESib SoTe. Ra (54), Ro, So.

Carex paniculata Greater Tussock-sedge; Seisg Bhadanach Mhòr
Boggy moorland, Baravaig, Sleat. Old records in 23, 24, 32 never confirmed. Locally common on Muck. Rare. Euro Temp. Skye (23, 24, 32, 61), M.

C. diandra Lesser Tussock-sedge; Seisg Bhadanach Bheag
Wet ground at edge of Loch Fada, 44; Loch Cleat, 47; with *C. paniculata* on Muck. Rare. Circ BoTe. Skye (36, 44, 47, 51), M.

C. otrubae False Fox-sedge; Seisg Gharbh Uaine
Wet ground on or near seashores. Rare. ESib SoTe. Skye (14, 24-26, 31, 34, 36, 47, 50, 51, 60), Ra (53), Ro, Ru, E, M.

C. arenaria Sand Sedge; Seisg Ghainmhich
Limited in Skye by lack of dune areas. Fiskavaig; Glen Brittle beach; Camasunary. Also recorded from 24 - where? Local. Euro Temp. Skye (24, 33, 42, 51), Ru, E, M, C, So.

C. remota Remote Sedge; Seisg Sgarta
Damp woodland. Portree; Tokavaig; and elsewhere. Very local, commoner in S Skye. Euro Temp. Skye (25, 35, 43, 44, 50, 51, 54, 60-62, 71), Ra (53, 54), Ru, E.

C. ovalis Oval Sedge; Seisg Ughach
Rough grassland. Locally common. ESib BoTe. Skye (all squares), Ra (53-55, 64, 65), Ro, Ru, E, M, C, Sc, So.

C. echinata Star Sedge; Seisg Reultach
Wet moorland and bogs. Common. Euro BoTe. Skye (all squares), Ra (53-55, 64, 65), Ro, Ru, E, M, C, Sc, So.

C. dioica Dioecious Sedge; Seisg Aon-cheannach
Base-rich flushes; rock ledges on hills. Local. Circ BAMo. Skye (all squares), Ra (53-55, 64, 65), Ro, Ru, E, M, C, Sc, So.

C. curta White Sedge; Seisg Bhàn
Edges of lochs and pools; bogs. Rare - or overlooked? Circ BoMo. Skye (14, 15, 24, 25, 32-34, 36, 42-45, 47, 50-52, 55, 56, 61, 62, 71, 72), Ra (53, 54), E, M, So.

C. hirta Hairy Sedge; Seisg Ghiobach
Roadside verge, Armadale (1980s), Rum (1997). Euro Temp. Skye (60), Ru.

C. lasiocarpa Slender Sedge; Seisg Choilleanta
Edges of lochs, or in wet ditches. Rare. Circ BoMo. Skye (31, 33, 43, 44, 46, 50, 61, 62, 71, 72), Ra (54), Ro, Ru.

C. rostrata Bottle Sedge; Seisg Shearragach
Lochs, bog pools, and ditches. Common. Circ BoTe. Skye (all squares), Ra (53-55, 64), Ro, Ru, E, M, C, Sc, So.

C. vesicaria Bladder-sedge; Seisg Bhalaganach *
N of Barkeval, Rum. Rare. Circ BoTe. Old records Skye, Ra, E, M.

C. sylvatica Wood-sedge; Seisg Choille
One record from grassy shore, others from woods or scrub. Very local, commoner in S Skye. EAsi Temp. Skye (23, 25, 26, 35, 43-45, 50-52, 60-62, 71, 72), Ra (53, 54), Ru, E, C.

C. flacca Glaucous Sedge; Seisg Liath-ghorm
Both dry and wet habitats, from near the sea to hills inland. Local. Euro SoTe. Skye (all squares), Ra (53-55, 64, 65), Ro, Ru, E, M, C, Sc, So.

C. panicea Carnation Sedge; Seisg a'Chruithneachd
Wet grassy places, moors and hills. Common. Euro BoTe. Skye (all squares), Ra (53-55, 64, 65), Ro, Ru, E, M, C, Sc, So.

C. laevigata Smooth-stalked Sedge; Seisg Mhìn
Harlosh Island; woodland at Ord; Raasay; Allt Camas na Geadaig, Scalpay. Rare. Ocea Temp. Skye (23, 60, 61, 71, 72), Ra (53-55, 64, 65), Ro, Sc, So.

C. binervis Green-ribbed Sedge; Seisg Fhèith-ghuirm
Moors and rough grassland. Common. Ocea Temp. Skye (all squares), Ra (53-55, 64, 65), Ro, Ru, E, M, C, Sc, So.

C. distans Distant Sedge; Seisg Fhada-mach
Salt-marshes and rocks near the sea. Local. Euro SoTe. Skye (14, 15, 23, 34, 42, 45, 47, 50, 62), Ra (53), Ru, E, M, C, Sc.

C. extensa Long-bracted Sedge; Seisg Anainn
Salt-marshes. Probably under-recorded. Local. Euro SoTe. Skye (14, 24, 25, 33, 34, 43-46, 52, 60-62, 71), Ra (53, 54), Ru, Ro, E, C.

C. hostiana Tawny Sedge; Seisg Odhar
Base-rich flushes. Local. Euro Temp. Skye (all squares except 54, 60), Ra (53-55, 64, 65), Ro, Ru, E, M, C, Sc.

C. viridula ssp. **brachyrrhyncha** *C. lepidocarpa* Long-stalked Yellow-sedge; Seisg Bhuidhe Fhad-chuiseagach
Wet places on base-rich soils. Occasional. Circ BoTe. Skye (14, 15, 23, 24, 26, 33, 35, 36, 42-45, 47, 51, 52, 54-56, 60-62, 71), Ra (53, 54, 64), Ro, Ru, E, M, C, Sc, So.

C. viridula ssp. **oedocarpa** *C. demissa* Common Yellow-sedge; Seisg Bhuidhe Choitcheann
Moors, hills, stony edges of lochs, etc. Locally common. Circ BoTe. Skye (all squares), Ra (53-55, 64, 65), Ro, Ru, E, M, C, So.

C. viridula ssp. **viridula** *C. serotina* Small-fruited Yellow-sedge; Seisg nam Measan Beaga
Damp places near the sea. Rather local, possibly under-recorded. Circ BoTe. Skye (15, 24, 34, 35, 41, 44, 45, 47, 51, 52, 56, 62, 71, 72), Ra (53, 55, 65), Ro, Ru, M, So.

C. viridula ssp. **viridula** var. **pulchella** *C. scandinavica*
Edges of runnels in salt-marshes. Probably under-recorded. Local. Circ BoTe. Skye (24, 35, 45, 52, 62, 71, 72).

C. pallescens Pale Sedge; Seisg Gheal
Wet grassland, and in woods (Raasay and Scalpay). Local. ESib BoTe. Skye (all squares except 37, 47), Ra (53-55, 64), Ro, Ru, E, M, C, Sc, So.

C. caryophyllea Spring Sedge; Seisg an Earraich
Dry calcareous grassland, Uig; Duntulm; Torrin. First sedge to flower, in May. Rare. ESib Temp. Skye (36, 41, 47, 50-52, 54, 62), Ra (53), Ru, E, M, Sc, So.

C. pilulifera Pill Sedge; Seisg Lùbach
Rough grassland, moors. Locally common. Euro Temp. Skye (all squares), Ra (53-55, 64), Ro, Ru, E, M, C, Sc, So.

C. limosa Bog-sedge; Seisg na Mòna
First recorded from Loch an Eilean, Sligachan; now known from wet, muddy loch margins all over Skye. Rare. Circ BoMo. Skye (23, 25, 31, 32, 34, 41-46, 50, 51, 55, 61, 62, 71, 72), Ra (53, 54, 64), Ru, Sc, So.

C. aquatilis Water Sedge; Seisg Uisge
Edge of Skeabost river, 1984. Circ BAMo. Skye (44).

C. nigra Common Sedge; Gainnisg
Wet grassland, moors and bogs. Common. ESib BoTe. Skye (all squares), Ra (53-55, 64, 65), Ro, Ru, E, M, C, Sc, So.

C. bigelowii Stiff Sedge; Dùr-sheisg
Damp stony places on hills; mountain grassland. Healaval Mhor; Cuillin hills; Storr-Quiraing ridge; Sgurr na Coinnich; etc on Skye. Locally common. Circ ArMo. Skye (23-25, 32, 42, 45-47, 52, 53, 71, 72), Ra (53), Ru.

C. pauciflora Few-flowered Sedge; Seis nan Lusan Gann
Wet moors and bogs. Difficult to identify when not in flower, and may be under-recorded. Commoner in S Skye. Circ BoMo. Skye (24, 32, 34, 35, 42-46, 50-53, 61, 62), Ra (53, 54, 64), Ru, Sc.

C. rupestris Rock Sedge; Seisg na Creige
Rock ledges on Ben Suardal (limestone) c. 800'-900', 1981. Circ ArMo. Skye (62).

C. pulicaris Flea Sedge; Seisg na Deargainn
Damp grassland, base-rich flushes, and rock ledges on hills. Common. SubO Temp. Skye (all squares), Ra (53-55, 64, 65), Ro, Ru, E, M, C, Sc, So.

Nardus stricta Mat-grass; Riasg
Moors and mountains. Common. Euro BoTe. Skye (all squares), Ra (53-55, 64, 65), Ro, Ru, E, M, C, Sc, So.

Festuca pratensis Meadow Fescue; Feisd *
N and W Skye mostly so far. Rare - or overlooked? Introduced? ESib BoTe. Skye (14, 24, 25, 35, 53, 60), Ru, M.

F. arundinacea Tall Fescue; Feisd Ard *
Roadside, Kilbride; Kinloch. Rare. ESib SoTe. Skye (24, 32, 52, 55, 71).

F. gigantea Giant Fescue; Feisd Mòr
Path to shore, Torrin, 52. Rare. Euro Temp. Skye (52, 55, 60-62, 71), Ra (54).

F. rubra Red Fescue; Feisd Ruadh
Grassland, salt-marshes, mountains. Check whether **F. arenaria** (*F. rubra* ssp. *arenaria*) occurs. Common. Circ WiBo. Skye (all squares), Ra (53-55, 64, 65), Ro, Ru, E, M, C, Sc, So.

F. ovina agg. Sheep's-fescue; Feur Chaorach
Dry grassland. Check for ssp. **ovina** and **F. filiformis** (= *F. ovina* ssp. *tenuifolia*). Common. EAsi BoTe. Aggregate in all Skye squares, Ra (53, 54, 65), Ro, Ru, E, M, C, Sc, So.

F. vivipara Viviparous Sheep's-fescue; Feur Chaorach Bèo-breitheach
Moors and hills. Common. Circ BAMo. Skye (all squares), Ra (53-55, 64, 65), Ro, Ru, E, M, C, Sc, So.

F. filiformis *F. tenuifolia, F. ovina* ssp. *tenuifolia* Fine-leaved
Sheep's-fescue; Mìn-fheur Chaorach
Grassy places. Skye (24, 52, 71), Ra (53, 54), Ru, M, Sc.

Lolium perenne Perennial Rye-grass; Seagal
Hayfields and waste places. Common. Euro SoTe. Skye (all
squares except 23, 41), Ra (53-55, 65), Ro, Ru, E, M, C, Sc, So.

L. multiflorum Italian Rye-grass; Breòillean Eadailteach
Lynedale, 35; also 46, 47, 52, 72 (Skye). Introduced. Circ
WiBo. Ra (53), E, M, C.

Vulpia bromoides Squirreltail Fescue, Barren Fescue; Feisd
Aimrod
Dry paths, roadsides, waste ground. Rare. SubM SubA. Skye
(14, 15, 24, 44, 46, 51, 55, 60, 71, 72), Ra (53-55, 64), Ro, E,
Sc, So.

Cynosurus cristatus Crested Dog's-tail; Coin-fheur
Grassland and cultivated ground. Common. Euro Temp. Skye
(all squares), Ra (53-55, 64, 65), Ro, Ru, E, M, C, Sc, So.

Puccinellia maritima Common Saltmarsh-grass; Feur Rèisg
Ghoirt
Salt-marshes. Locally common. EAsi BoTe. Skye (23-26,
32-35, 43-45, 50-52, 56, 60-62, 71, 72), Ra (53, 54), Ro, Ru,
E, M, C, Sc, So.

Briza media Quaking Grass; Conan Cumanta
Grassland on base-rich soils. Rare. Euro Temp. Old records all
N Skye. Refound, Glen Conon (46) 1991. Raasay, 1987. Skye
(26, 34, 46, 47), Ra (53).

Poa annua Annual Meadow-grass; Tràthach Bliadhnail
Cultivated ground, gardens, moorland and hills. Common.
ESib WiTe. Skye (all squares), Ra (53-55, 64, 65), Ro, Ru, E, M,
C, Sc, So.

P. trivialis Rough Meadow-grass; Tràthach Garbh
Grassland, waste places. Common. ESib WiTe. Skye (all
squares except 47), Ra (53-55, 64, 65), Ro, Ru, E, M, C, Sc, So.

P. humilis (*P. subcaerulea*) Spreading Meadow-grass; Tràthach
Sgaoilte
Damp grassland, including hills. Other records may have been
included in *P. pratensis* agg. Local. Skye (14, 25, 36, 37, 42-46,
50-53, 55, 61, 62, 71, 72), Ro, Ru, M.

P. pratensis agg. Smooth Meadow-grass; Tràthach Mìn
Grassland. Perhaps still under-recorded. Circ WiTe. Skye (all squares except 34, 37, 41, 44, 53-55, 71), Ra (53-55, 64, 65), Ro, Ru, E, M, C, Sc, So.

P. compressa Flattened Meadow-grass; Tràthach na Duilleige Leathainn
Wall at Ardvasar, 60 (Skye). Doubtfully native, but more likely so on Rum. Euro Temp. Skye (60), Ru.

P. palustris Swamp Meadow-grass; Tràthach Lèana
Marshy ground on Canna. Record from 1930s re-found 1995. Dalavil, Skye 1998. Skye (50), C.

P. glauca *P. balfourii* Glaucous Meadow-grass; Tràthach Liath-ghorm
Mountain rock ledges. Cuillin (rare); Storr-Quiraing ridge (frequent). Circ BAMo. Skye (42, 44-47, 54, 55), Ru.

P. nemoralis Wood Meadow-grass; Tràthach Coille
Several of these records may include the taxon *P. balfourii* that is now considered to be synonymous with *P. glauca* by Stace (1977). Grassy bank of Bealach Beag burn, Storr; Ben Edra; Sgurr Mor. Rare. Circ BoTe. Skye (24,35, 36, 44-47, 50, 52, 60, 61, 71, 72), Ro, Ru, E, C.

P. alpina Alpine Meadow-grass; Tràthach Ailpeach
Rock ledges. Storr-Quiraing ridge (occasional); Cuillin corries (very rare). Circ ArMo. Skye (42, 45, 46, 55), Ru.

Dactylis glomerata Cock's-foot; Garbh-fheur
Grassland, waste places. Common. ESib SoTe. Skye (all squares), Ra (53, 54), Ru, E, M, C, Sc, So.

Catabrosa aquatica Whorl-grass: Feur-sùghmhor
In wet sand or shingle near the sea, not always in the same place. Rare. Euro BoTe. Skye (24, 32, 42, 47, 51, 56, 72), Ru, E, M, C.

Catapodium marinum Sea Fern-grass, Stiff Sand-grass; Feur Gainmhich
Salt-marsh at Caolas Scalpay, on Skye; Kilmory, Rum; Laig, Eigg. Med Atl. Skye (52), Ru, E.

Glyceria fluitans Floating Sweet-grass: Mìlsean Uisge
Shallow streams, pools and ditches. Local. Euro Temp. Skye (14, 15, 24-26, 34-37, 42-47, 50-52, 55, 60-62, 71, 72), Ra (53-55), Ro, Ru, E, M, C, Sc, So.

G. declinata Small Sweet-grass; Milsean Uisge Beag
Less common than above - or under-recorded. SubO Temp. Skye (14, 15, 25, 26, 32, 33, 43, 45, 47, 51, 53, 55, 61, 62, 71), Ru, E, M, Sc, So.

Melica nutans Mountain Melick; Meilig an t-Sleibhe Critheanach
On limestone at Tokavaig; Suardal; on ultrabasic rocks at Geary. Rare. EAsi BoTe. Skye (26, 43, 61, 62).

Helictotrichon pubescens Downy Oat-grass; Feur Coirce Clumhach
Base-rich grassland. Locally common. Euro Temp. Skye (14, 15, 23-26, 32-36, 42-47, 50-52, 54-56, 60-62, 71, 72), Ra (53, 54), Ru, E, M, C.

H. pratense Meadow Oat-grass; Feur Ccoirce Lòin *
Most likely to be found on limestone. Rare. Euro Temp. Skye (24, 26, 46, 61, 62).

Arrhenatherum elatius False Oat-grass; Feur Coirce Brèige
In cultivated ground, often in form *bulbosum* (Onion Couch). Locally too plentiful. Euro Temp. Skye (all squares), Ra (53, 54, 65), Ro, Ru, E, M, C, Sc, So.

Trisetum flavescens Yellow Oat-grass; Feur Coirce Buidhe
Roadside Elgol (1960s) 51; also 24 (Skye). Rare - or Introduced? Euro Temp. Ru, E. All old records.

Koeleria macrantha (*K. cristata*) Crested Hair-grass; Cuiseag Dhosach
Grassland near the sea. All records from west side of Skye. Missing or overlooked elsewhere? Rare. Circ Temp. Skye (14, 23-26, 31-34, 41, 50, 56), Ru, E, M, C.

Deschampsia cespitosa Tufted Hair-grass; Cuiseag Airgid
Rough grassland, including hills. Common. Circ WiBo. Skye (all squares), Ra (53-55, 64), Ro, Ru, E, M, C, Sc.

D. cespitosa ssp. **alpina** *D. alpina* Alpine Hair-grass; Mòin-fheur Ailpeach
Rocks and scree on mountains, usually viviparous. Cuillin hills; Storr; Ben Edra; Blaven; Red Hills. Local. Skye (42, 45, 46, 52), Ru.

D. setacea Bog Hair-grass; Mòin-fheur Bogaich
Boggy edges of lochs. Loch an Eilean, Sligachan; Loch Buidhe, NW of Heast; Lochain Dubha, by Broadford-Armadale road; also Raasay. Rare. Ocea Temp. Skye (42, 43, 50, 51, 61, 62), Ra (53, 54).

D. flexuosa Wavy Hair-grass; Mòin-fheur
  Moorland and grassland. Common. Euro BoTe. Skye (all squares), Ra (53-55, 64, 65), Ro, Ru, E, M, C, Sc, So.
Holcus lanatus Yorkshire-fog; Feur a'Chinn Bhàin
  Poor hayfields, waste ground. Common. Euro SoTe. Skye (all squares), Ra (53-55, 64, 65), Ro, Ru, E, M, C, Sc, So.
H. mollis Creeping Soft-grass; Mìn-fheur
  Woods and wet moorland. Locally common. Euro Temp. Skye (all squares except 34, 41, 42), Ra (53-55, 64, 65), Ro, Ru, E, M, C, Sc, So.
Aira caryophyllea Silver Hair-grass; Sìdh-fheur
  Dry places and on walls. Less common than *A. praecox*, perhaps still under-recorded. Euro SoTe. Skye (14, 15, 24-26, 32-35, 42, 44, 45, 50, 51, 53-56, 60-62, 72), Ra (53, 54), Ru, E, M, C.
A. caryophyllea ssp. **multiculmis**
  Introduced. Skye (25, 51, 62, 71, 72).
A. praecox Early Hair-grass; Cuiseag an Earraich
  On turf dykes and dry bare ground, in spring. Locally common. SubO SoTe. Skye (all squares), Ra (53-55, 64, 65), Ro, Ru, E, M, C, Sc, So.
Anthoxanthum odoratum Sweet Vernal-grass; Borrach
  Grassland, moors, and hills. First grass to flower. Common. ESib WiTe. Skye (all squares), Ra (53-55, 64, 65), Ro, Ru, E, M, C, Sc, So.
Phalaris arundinacea Reed Canary-grass; Cuiseagrach
  Wet ground, and margins of pools and ditches. Dead leaves last the winter, unlike *Phragmites*. Locally plentiful. Circ BoTe. Skye (14, 15, 23-26, 32, 33, 36, 37, 43-47, 50-53, 60-62, 71, 72), Ra (54), Ru, E, M, C.
Agrostis capillaris *A. tenuis* Common Bent; Freothainn
  Acid grassland. Common. ESib BoTe. Skye (all squares), Ra (53-55, 64, 65), Ro, Ru, E, M, C, Sc, So.
A. gigantea Black Bent; Fioran Dubh
  Cultivated ground, woods. Portree, 44; Prabost, 45; Heast, 61; Dunringell, Kyleakin, 72 (all Skye). Perhaps elsewhere? No records from other islands. Rare. EAsi SoTe. Skye (44, 45, 61, 71, 72).
A. stolonifera Creeping Bent; Fioran
  Waste ground; salt-marshes. Locally plentiful. Circ WiTe. Skye (all squares), Ra (53-55, 64, 65), Ro, Ru, E, M, C, Sc, So.

A. canina Velvet Bent; Fioran Mìn

> Dry acid grassland. Needs separating into ssp. **canina** and ssp. **montana**. Common. Circ BoTe. Skye (all squares except 37, 47, 54), Ra (53-55, 64, 65), Ro, Ru, E, M, C, Sc, So.

Calamagrostis epigejos Wood Small-reed, Bush-grass; Cuilc-fheur Coille

> Grassy slopes or rock crevices, on Jurassic rocks or limestone. An Leac; Rigg; Tokavaig; Suardal; Raasay. Rare. EAsi BoTe. Skye (41, 55, 61, 62), Ra (53-55), Ro, E.

Ammophila arenaria Marram; Muran

> Glen Brittle beach, 42. Also old records (error?) in 37, 47 (Skye). Rare. Euro SoTe. Skye (37, 42, 47), Ru, E, M, C.

Alopecurus pratensis Meadow Foxtail; Fiteag an Lòin

> Cultivated grassland, road verges. Locally common. ESib BoTe. Skye (14, 15, 24-26, 32-37, 44-47, 50-52, 60-62), Ra (53, 54), Ru, E, M, C, Sc.

A. geniculatus Marsh Foxtail; Fiteag Cham

> Wet ground. Local. Euro BoTe. Skye (all squares except 23, 41, 54), Ra (53-55), Ro, Ru, E, M, C, Sc, So.

Phleum pratense Timothy; Feur Cait

> Scattered specimens, roadsides and waste ground. Records include **P. bertolonii** Skye (14, 15, 42-45), Ra (53, 54). Rare. Probably introduced. ESib Temp. Skye (14, 15, 24, 26, 32, 33, 42-45, 47, 50-52, 61, 62, 71, 72), Ra (53, 54), Ru, E, C.

Bromus hordaceus agg. Soft-brome; Bromas Bog

> Grassland and waste ground. Occasional. Euro SoTe. Skye (14, 15, 24-26, 33-37, 44, 45, 47, 50-53, 55, 56, 60, 61, 71, 72), Ra (53, 54), Ru, E, M, C, So.

B. lepidus Slender Soft-brome; Bromas Bog Caol *

> Old records (1950s) Skye (24) and Ra (53) only. Casual? Introduced. Skye (24), Ra (53).

Bromopsis ramosa *Bromus ramosus, Zerna ramosa* Hairy-brome; Bromas Giobach

> In woodland or scrub; sometimes on grassy slopes below coastal cliffs. Rare. Euro Temp. Skye (25, 26, 33, 43, 44, 50, 51, 53-55, 61, 62, 71, 72), Ra (53, 54), E.

Anisantha sterilis *Bromus sterilis* Barren Brome; Bromas Aimrid *

> Walls, waste ground. Raasay and Skye. Introduced? Euro SoTe. Skye (60), Ra (53).

Brachypodium sylvaticum False-brome; Bromas Brèige
  Woodland and scrub. Locally common. Euro Temp. Skye (all squares), Ra (53-55, 64, 65), Ro, Ru, E, M, C, Sc, So.
Elymus caninus Bearded Couch; Taithean *
  Woods. Uncommon, or else overlooked. Old 1950s records require confirmation. ESib BoTe. Skye (14, 32, 33, 36, 51, 52, 61, 62), Ra (54).
Elytrigia repens *Agropyron repens* Common Couch; Feur a'Phuint
  An infuriating weed of cultivated ground; probably still under-recorded. ESib WiTe. Skye (14, 15, 24-26, 33, 34, 36, 37, 42, 44, 45, 50-53, 60-62, 71, 72), Ra (53, 54), Ru, E, M, C, So.
E. repens x **E. juncea** = **E.** x **laxa**
  Churchton Bay, Raasay, 1997. Ra (53).
E. juncea ssp. **boreoatlantica** Sand Couch; Glas-fheur
  Rare in Skye from lack of suitable habitat. Kilbride Point; Glen Brittle beach; Camas Croise, Isle Ornsay. Euro SoTe. Skye (36, 42, 61), Ru, E, M, C.
Leymus arenarius Lyme-grass; Taithean
  Kilmaluag; Rigg; in both places growing in shingle. Old records for 45, 50 unconfirmed. Euro BAMo. Skye (45, 47, 50, 55), Ra (53), E (Introduced).
Danthonia decumbens *Sieglingia decumbens* Heath-grass; Feur Monaidh
  Moorland; hill grassland. Common. Euro Temp. Skye (all squares), Ra (53-55, 64, 65), Ro, Ru, E, M, C, Sc, So.
Molinia caerulea Purple Moor-grass; Fianach
  Wet moorland and mountains. Common. ESib BoTe. Skye (all squares), Ra (53-55, 64, 65), Ro, Ru, E, M, C, Sc, So.
Phragmites australis *P. communis* Common Reed; Cuilc
  In lochs, ditches and other wet places. Local. Circ WiTe. Skye (14, 15, 24-26, 32-37, 41-47, 50, 52, 53, 60-62, 71, 72), Ra (53-55), Ro, Ru, E, M, C, Sc, So.
Sparganium erectum Branched Bur-reed; Seisg Rìgh
  Burn at Monkstadt; edge of R. Gremiscaig; burn joining Kilmaluag R.; ditch at Aird of Sleat; Allt Port na Cullaidh, Elgol. Rare. Circ Temp. Skye (25, 36, 37, 46, 47, 50, 51, 61), M.
S. emersum Unbranched Bur-reed; Seisg Rìgh Madaidh *
  Lochs and ditches. Rare. Circ BoTe. Check with other *Sparganium* spp. 35, 62 (Skye). Old records in Ra (54, 55), E, Sc, So.

S. angustifolium Floating Bur-reed; Seisg Rìgh air Bhog
Peaty lochs in hill areas. Locally plentiful. Euro BoMo. Skye (14, 15, 23-26, 33, 34, 41-47, 51, 52, 55, 56, 61, 71), Ra (53-55, 64, 65), Ro, Ru, Sc, So.

S. natans (*S. miminum*) Least Bur-reed; Seisg Rìgh Mion
Lochs and pools. Less common than above. Most records from S Skye - overlooked elsewhere? Circ BoTe. Skye (23, 35, 42, 43, 50-52, 61, 62, 71), Ra (53, 54), Ro, E, M, Sc, So.

Typha latifolia Bulrush; Cuigeal nam Ban-sìdh
Cleadale, Eigg, 2000. Circ SoTe. E.

T. angustifolia Lesser Bulrush; Bodan
Loch a'Mhuilinn, Scalpay. Old record from Arish Burn, Raasay, may have been this and not *T. latifolia*. Scalpay plants probably introduced. ESib Temp. Ra (53), Sc.

Tofieldia pusilla Scottish Asphodel; Bliochan Albannach
Recorded from "East corrie, Sgurr nan Eag" in 1947, and confirmed from top of An Garbh Choire (same area) in 1978. Rare. Easier to find on Rum (Barkeval). Circ ArMo. Skye (41), Ru.

Narthecium ossifragum Bog Asphodel; Fianach
Wet moorland. Common. Ocea BoTe. Skye (all squares), Ra (53-55, 64, 65), Ro, Ru, E, M, C, Sc, So.

Paris quadrifolia Herb-Paris; Aon-dhearc
Boulder scree, below Sgurr a'Bhagh (1960s); ravine of Allt a'Ghlinne; Tokavaig gorge, Sleat; limestone grikes at Suardal and Corrychatachan; the first two not on limestone, but probably ultrabasic. Rare. ESib BoTe. Skye (25, 26, 61, 62). Old record from Scalpay (1930s).

Scilla verna Spring Squill; Lear-uinnean
Dry grassy places on the coast. Reported as seen on Skye, but never confirmed. Nearest is on Canna. Ocea Temp. C.

Hyacinthoides non-scripta (*Endymion non-scriptus*) Bluebell, Wild Hyacinth; Bròg na Cuthaig
More often in open grassland than in woodland. Rock ledges on Sgurr an Fheadain (Cuillin) at c. 1200ft. Locally plentiful. Ocea Temp. Skye (all squares), Ra (53-55, 64, 65), Ro, Ru, E, M, C, Sc, So.

Allium ursinum Ramsons, Wild Garlic; Creamh
Woodland, and in boulder scree; also often on sea cliffs. Locally common. Euro Temp. Skye (all squares), Ra (53, 54), Ru, E, M, C, Sc, So.

A. vineale Wild Onion; Gairleag Moire
Isolated rock ledges in NE Raasay, 1991. Euro Temp. Ra (54).

Ruscus aculeatus Butcher's-broom; Calg-bhealaidh
Ruined garden at Fiskavaig; Raasay House grounds. Introduced. SubM SubA. Skye (33), Ra (53).

Iris pseudacorus Yellow Iris, Yellow Flag; Seileasdair
Wet places, including seashores. Locally common. Euro SoTe. Skye (all squares), Ra (53, 54), Ro, Ru, E, M, C, Sc, So.

Crocosmia x **crocosmiflora** *C. aurea* x *C. pottsii* Montbretia
Naturalised - an escape from garden or estate plantings. Skye (14, 15, 24-26, 33-35, 42-46, 50-52, 60, 61, 71, 72), Ra (53, 54), M, C, So.

Cephalanthera longifolia Narrow-leaved Helleborine; Eileabor Geal
In hazel scrub at Calligarry, Sleat; also Achnacloich; Dalavil wood. Rare. Euro Temp. Skye (50, 60).

Epipactis atrorubens Dark-red Helleborine; Eileabor Dearg
In limestone 'grikes', Allt nan Leac, S. of Camas Malag; Leac nan Craobh, S of Torrin; Tokavaig; Suardal; and on Raasay. Often plants without flowers. Rare. ESib BoTe. Skye (51, 52, 55, 61, 62), Ra (53, 54).

E. helleborine Broad Helleborine; Eileabor Leathann
Half a dozen scattered plants on limestone, south of Torrin. Rare. EAsi Temp. Skye (51, 52).

Neottia nidus-avis Bird's-nest Orchid; Mogairlean Nead an Eòin
Woodland above Fearns-Hallaig path, Raasay, 1969. Not seen recently. Scorrybreck, Portree, 1997. Rare. ESib Temp. Skye (44), Ra (53).

Listera ovata Common Twayblade; Dà-dhuilleach Coitcheann
Open grassland and in woods; limestone 'grikes' and rock ledges at Suardal and elsewhere; on limestone only, Raasay. Very local. ESib BoTe. Skye (15, 24, 26, 33, 36, 45, 46, 51-53, 55, 56, 60-62), Ra (53, 54), E.

L. cordata Lesser Twayblade; Dà-dhuilleach Monaidh
Much more plentiful than *L. ovata*. Often overlooked under damp heather; also in *Sphagnum* on damp banks. Begins flowering in May. Local. Circ BoMo. Skye (all squares except 32, 37, 41, 56), Ra (53-55, 64, 65), Ro, Ru, E, Sc, So.

Hammarbya paludosa Bog Orchid; Mogairlean Bogaich
Boggy ground, Loch an Eilean, Sligachan; near Loch Ainort; near Lochan Fada; N end of Raasay etc. Rare and difficult to find if not in flower. Circ BoMo. Skye (31, 41-43, 52, 61, 72), Ra (54, 65), Ru, E, Sc, So.

Platanthera chlorantha Greater Butterfly-orchid; Mogairlean an Dealain-dè Mòr
Rough grassland. Numbers vary from one year to another, possibly reflecting weather differences. Local. Euro Temp. Skye (all squares except 23, 37, 41, 54, 55, 72), Ra (53-54), Ru, E, C, Sc, So.

P. bifolia Lesser Butterfly-orchid; Mogairlean an Dealain-dè Beag
Often in same stretch of ground as above. Local. EAsi BoTe. Skye (24, 33-36, 41-47, 50-54, 56, 60-62, 71, 72), Ra (53-55, 64), Ro, Ru, E, M, C, Sc, So.

Pseudorchis albida *Leucorchis albida* Small-white Orchid; Mogairlean Bàn Beag
Scattered plants on dry moorland or short turf. Rare. Euro BoMo. Skye (14, 15, 24-26, 32, 33, 35, 36, 41-45, 47, 50-53, 55, 56, 60-62, 72), Ra (53, 54), Ru, E.

Gymnadenia conopsea Fragrant Orchid; Lus Taghte
Base-rich grassland, but sometimes among heather. Locally common. The Skye plants match the recent description of ssp. **borealis** but ssp. **densiflora** might also occur. EAsi BoTe. Skye (all squares except 23, 32, 34, 72), Ra (53, 54, 64), Ro, Ru, E, M, C.

Coeloglossum viride Frog Orchid; Mogairlean Losgainn
Records suggest it prefers limestone or Jurassic areas to basalt. Very local. Circ BoMo. 24-26, 33, 35-37, 46, 47, 51, 52, 54-56, 61, 62 (Skye). Ra (53, 54), Ru, E, M, C, Sc.

Dactylorhiza fuchsii Common Spotted-orchid; Urach-bhallach
Damp grassland. Check on possible occurrence on Skye of ssp. **hebridensis** (Rum). Locally common. ESib Temp. Skye (all squares except 41, 53), Ra (53-55), Ro, Ru, E, M, C, Sc.

D. maculata ssp. **ericetorum** Heath Spotted-orchid; Mogairlean Mòinteach
Moorland and hills. Common. ESib BoTe. Skye (all squares), Ra (53-55, 64, 65), Ro, Ru, E, M, C, Sc, So.

D. incarnata Early Marsh-orchid; Mogairlean Lèana

Wet peaty ground. Records all ssp. **incarnata** (flowers flesh pink). Very local. ESib BoTe. Skye (14, 15, 23-25, 33-35, 42-45, 47, 50-53, 60-62, 71, 72), Ra (53, 54), Ro, Ru, E, M, C, Sc, So.

Ssp. **pulchella** (flowers magenta). Skye (36, 42-44, 51, 53, 61, 62, 72), Ra (53, 54, 64), Ro, Ru, M. Ssp. **coccinea** found near Ord, Sleat, 1990. Skye (61); Kilmory, Rum; Cleadale, Eigg, 1995; Eilean nan Each, Muck, 1996, Canna 2001; refound Raasay (1951 site) 1997. Skye (61), Ra (53), Ru, E, M, C.

D. purpurella Northern Marsh-orchid; Mogairlean Purpaidh

Wet grassland and marshy ground. Locally common. Ocea BoMo. Skye (all squares except 23, 37, 41, 54), Ra (53, 54), Ro, Ru, E, M, C, Sc, So.

D. lapponica Lapland Marsh Orchid; Mogairlean Laplainneach

New to British Flora, 1988. Base-rich flushes, Rum, 1989. Raasay, 1994. Skye, 2002. Euro BoMo. Skye (52), Ra (54), Ru.

Hybrids. The following have been found in Skye, Raasay, or Rum, and have been determined by experts - **Dactylorhiza maculata** ssp. **ericetorum** x **D. purpurella**; **D. fuchsii** x **D. purpurella**; **Gymnadenia conopsea** x **D. purpurella**; **G. conopsea** x **D. fuchsii**; **D. incarnata** x **D. purpurella**; **Pseudorchis albida** x **D. maculata**. The first is the commonest. More work is needed on the identification and recording of these orchid hybrids.

Orchis mascula Early-purple Orchid; Saighdear Dearg

Occurs in patches in grassland; singly on sea-cliff and mountain ledges. First orchid to flower, in May. Locally common. Euro Temp. Skye (all squares except 34, 50), Ra (53-55, 64), Ru, E, M, C.

ACKNOWLEDGEMENTS

We wish to thank everyone who has contributed plant records for v.c. 104 over the years, including anyone we have inadvertently missed from the list in Appendix III. We are particularly indebted to J. Bevan, H.H. Birks, P.F. Braithwaite, J.G. Brownlie, S.J. Bungard, R.H. Dobson, R.M. Dobson, M. Gregory, D.A. Ratcliffe, and J.G. Roger for their many important records and valuable insights into the flora of these islands.

We acknowledge the help we have received in determination of difficult or critical specimens by members of the Botanical Society of the British Isles; the Natural History Museum, London; Royal Botanic Gardens, Edinburgh; the former Botany School, University of Cambridge; and the Biological Records Centre, Monks Wood, in particular J. Bevan, A, Dudman, C.C. Haworth, D.J. McCosh, D. McKean, the late F.H. Perring, C.D. Preston, A.J. Richards, P.D. Sell, A.J. Silverside, S.M. Walters, the late C. West, and P.F. Yeo.

We are grateful to the late F.H. Perring for all his help in the production of the second (1980) edition, to D.A. Pearman for much valuable advice, and to C. Jenks for invaluable help in the preparation and production of this edition.

We greatly appreciate financial contributions from the Botanical Society of the British Isles, the Blodwen Lloyd Binns Trust of the Glasgow Natural History Society, and the University of Bergen, towards the costs of production and publication.

BOTANICAL SOCIETY OF THE BRITISH ISLES

The BSBI was founded in 1836 and has a membership of nearly 3000. It is the major source of information on the status and distribution of British and Irish flowering plants and ferns. The Society arranges conferences and field meetings throughout the British Isles and, occasionally, abroad. It organises plant distribution surveys and publishes plant atlases and handbooks on difficult groups of plants. It has a panel of referees available to members to name problematic specimens. It welcomes as members all botanists, professional and amateur alike. Details of membership and any other information about the Society may be obtained from: The Hon. General Secretary, Botanical Society of the British Isles, c/o Department of Botany, the Natural History Museum, Cromwell Road, London, SW7 5BD, or from www.bsbi.org.uk.

REFERENCES

Asprey, G.F. 1947. The vegetation of the islands of Canna and Sanday, Inverness-shire. *Journal of Ecology* **34**: 182-193.

Barkley, S.Y. 1953. The vegetation of the island of Soay, Inner Hebrides. *Transactions of the Botanical Society of Edinburgh* **36**: 119-131.

Birks, H.J.B. 1973 *Past and present vegetation of the Isle of Skye – A palaeoecological study.* Cambridge University Press, London.

Birks, H.J.B. & Birks, H.H. 1974. Studies on the bryophyte flora and vegetation of the Isle of Skye: I. Flora. *Journal of Bryology.* **8**: 19-64, 197-254.

Birks, H.J.B., Murray, C.W., Birks, H.H. & Murray, R.M. 1991. Notes on the flora and vegetation of Canna and Sanday. *Botanical Journal of Scotland.* **46**: 15-26.

Braithwaite, M. 2001. *Canna's Wild Flowers.* Privately published.

Bungard, S.J. 1993. *Raasay Plant List* (+ subsequent updates). Raasay Heritage Society, Isle of Raasay.

Butterfield, I. & Bell, S.L. 1996. Freshwater macrophyte survey of Skye and Lochalsh 1989. *Scottish Natural Heritage Research, Survey and Monitoring Report.* **54**: 27 pp plus 8 appendices.

Campbell, J.L. 1984. *Canna, The Story of a Hebridean Island.* Oxford University Press, Oxford.

Clark, J.W. & MacDonald, I. 1999. *Ainmean gaidhlig lusan – Gaelic Names of Plants.* J.W. Clark, Fort William.

Dobson. R.H. & Dobson, R.M. 1985. The Natural History of the Muck Islands, north Ebudes I: Introduction and Vegetation with a List of Vascular Plants. *Glasgow Naturalist* **21**: 13-28.

Eggeling, W.J. 1965. Check List of the Plants of Rhum, Inner Hebrides (v.c. 104 North Ebudes) Part 1. Stoneworts, Ferns and Flowering Plants. *Transactions of the Botanical Society of Edinburgh.* **40**: 20-59.

Farmer, A.M. 1984. Aquatic angiosperm communities from lochs on Rhum. *Transactions of the Botanical Society of Edinburgh.* **44**: 229-236.

Gilbert, O.L. 1984. Some effects of disturbance on the lichen flora of oceanic hazel woodland. *Lichenologist.* **16**: 21-30.

Magnusson, M. 1997. *Rum: Nature's Island.* Luath Press, Edinburgh.

Pearman, D.A. & Walker, K.J. 2004. An examination of J.W. Heslop Harrison's unconfirmed plant records from Rum. *Watsonia*. **25**: 45-63.

Preston, C.D. (ed.) 2004. John Raven's report on his visit to the Hebrides, 1948. *Watsonia*. **25**: 17-44.

Preston, C.D. & Hill, M.O. 1997. The geographical relationships of British and Irish plants. *Botanical Journal of the Linnean Society*. **124**: 1-120.

Ratcliffe, D.A. 1977. *A Nature Conservation Review*. Two volumes. Cambridge University Press, Cambridge.

Raven, J.A. 1949. Alien plant introductions on the Isle of Rhum. *Nature*. **163**: 104-105.

Sabbagh, K. 1999a. Rum's the word. *New Scientist* (7th August) 44-47.

Sabbagh, K. 1999b. *A Rum affair*. Allen Lane, The Penguin Press, London.

Stace, C.A. 1997. *New Flora of the British Isles* (Second edition). Cambridge University Press, Cambridge.

SUGGESTIONS FOR FURTHER READING

General

Boyd, J.M. & Boyd, I.L. 1990. *The Hebrides*. Collins, London.

Clutton-Brock, T.H. & Ball, M.E. (Eds.) 1987 *Rhum – the Natural History of an Island*. Edinburgh University Press, Edinburgh.

Love, J.A. 2001. *Rum – a landscape without figures*. Birlinn, Edinburgh.

Ratcliffe, D.A. 1977. *Highland Flora*. Highlands and Islands Development Board, Inverness.

Rixson, D. 2001. *The Small Isles – Canna, Rum, Eigg, and Muck*. Birlinn, Edinburgh.

Geology

Goodenough, K. & Bradwell, T. 2004. *Rum and the Small Isles – a landscape fashioned by geology*. Scottish Natural Heritage, Battleby.

Stephenson, D. & Merritt, J. 2002. *Skye – a landscape fashioned by geology*. Scottish Natural Heritage, Battleby.

Plant Identification

Blamey, M., Fitter, R. & Fitter, A. 2003. *Wild Flowers of Britain and Ireland.* A. & C. Black, London

Blamey, M. & Grey-Wilson, C. 2003 *Wild Flowers of Britain and Northern Europe.* Cassell, London.

Rose, F. 1981. *The Wild Flower Key.* Frederick Warne, London.

Rose, F. 1989. *Colour Identification Guide to the Grasses, Sedges, Rushes, and Ferns of the British Isles and north-western Europe.* Viking, London.

Plant Distribution and Status

Preston, C.D. & Corft, J.M. 1997. *Aquatic plants in Britain and Ireland.* Harley Books, Colchester.

Preston, C.D., Pearman, D.A. & Dines, T.D. 2002. *New Atlas of the British and Irish Flora.* Oxford University Press, Oxford.

Stewart, A., Pearman D.A. & Preston, C.D. 1994. *Scarce Plants in Britain.* Joint Nature Conservancy Committee, Peterborough.

Wiggington, M.J. 1999. *British Red Data Books 1: Vascular Plants.* Joint Nature Conservancy Committee, Peterborough.

Mountain Flora and Vegetation

Averis, A.M., Averis, A.B.G., Birks, H.J.B., Horsfield, D., Thompson, D.B.A. & Yeo, M.J.M. 2004. *An Illustrated Guide to British Upland Vegetation.* Joint Nature Conservancy Committee, Peterborough.

Raven, J. & Walters, M. 1956. *Mountain Flowers.* Collins, London.

Vegetation and Plant Ecology

Currie, A. & Murray, C.W. 1983. Flora and vegetation of the Inner Hebrides. *Proceedings of the Royal Society of Edinburgh* **83B**; 293-318.

Vegetational History

Birks, H.J.B. & Williams, W. 1983. The late-Quaternary vegetational history of the Inner Hebrides. *Proceedings of the Royal Society of Edinburgh* **83B**; 269-292.

APPENDIX I — Omitted records

The following records have been omitted, either on distributional grounds, or, though recorded at some time in the past 100 years, they have not been re-found by local or visiting botanists in the last 40 years or rooted specimens have not been found. Six species from the 1973 'omitted' list *(Lycopodiella inundata, Ranunculus auricomus, Moehringia trinervia, Vicia sativa* ssp. *nigra (V. angustifolia), Veronica hederifolia,* and *Hieracium schmidtii)* were re-instated in 1980, while 24 from the main list were relegated as doubtful. Since 1980, twelve more species have been re-instated *(Viola canina, Agrimonia odorata, Loiseleuria procumbens, Vaccinium microcarpon, Campanula rapunculoides, Potamogeton crispus, Woodsia alpina, Pilularia globulifera, Trifolium campestre, Polygonum lapathifolium,* and *Briza media),* while a further nine are now 'doubtful' or errors (marked as •).

	Adoxa moschatellina		*Galeopsis ladanum*
	Aethusa cynapium		*Galium pumilum*
	Alchemilla minor		*G. uliginosum*
	A. glomerulans		*Hieracium flocculosum*
	Andromeda polifolia	•	*H. longilobum*
	Anthemis nobilis	✶	*Hordeum murinum*
	Apium graveolens		*Hypericum maculatum*
✶	*Arum maculatum*		*Hypochoeris glabra*
	Avena strigosa	•	*Juncus capitatus*
	Callitriche polymorpha	•	*J. mutabilis*
	Carduus nutans		*Knautia arvensis*
	C. acanthoides		*Lamium album*
	Carex saxatilis		*Leontodon hispidus*
	C. acutiformis		*Myosotis brevifolia*
•	*C. paupercula*	•	*Myriophyllum spicatum*
+	*Carex* hybrids	✶	*Ophrys apifera*
	Carum carvi		*Peucedanum ostruthium*
	Cerastium alpinum		*Potamogeton epihydrus*
	Chelidonium majus		*P. pectinatus*
	Cicendia pusilla		*Primula veris*
✶	*Cichorium intybus*		*Ranunculus sardous*
•	*Circaea alpina*		*R. aquatilis*
	Conium maculatum		*R. reptans*
	Convolvulus arvensis		*Rosa tomentosa*
	Cystopteris montana		*Sagina normaniana*
	Eryngium maritimum		*Saxifraga tridactylites*

Scandix pecten-veneris
Sedum telephium
- Silene vulgaris
- Sinapis alba
- Thlaspi arvense

Trifolium aureum
Vaccinium uliginosum
+ Various hybrids
✱ Verbascum thapsus
Viola reichenbachiana

✱ Casual or introduced
+ Hybrids still to be checked
- 'Doubtful' or error

APPENDIX II — Estate and Forest Enterprise Woodland

These lists include information supplied by Duncan MacInnes (Clan Donald Centre, Armadale Castle), D.C. Maclean (Dunvegan Castle), and David Robertson (Head Forester, F.E. Skye) in 1980, but are probably not complete.

ESTATE WOODLAND - Trees

Abies alba
A. concolor
A. procera
Acer campestre
A. platanoides
A. pseudoplatanus
Aesculus hippocastanum
A. x carnea
Araucaria araucana
Betula spp.
Castanea sativa
Cedrus spp.
Chamaecyparis lawsoniana
Cryptomeria japonica
Cupressus macrocarpa
Fagus sylvatica
Fraxinus excelsior
Ilex spp.
Juglans regia
Larix spp.

Liquidambar styraciflua
Liriodendron tulipifera
Mespilus canadensis
Picea sitchensis
Pinus spp.
Populus alba
Prunus avium
Pseudotsuga japonica
P. menziesii
Quercus borealis
Q. robur
Salix spp.
Sciadopitys verticillata
Sequoiadendron giganteum
Sorbus spp.
Taxus baccata incl. var. *fastigiata*
Thuja plicata
Tilia x europea
Tsuga heterophylla
Ulmus procera

Shrubs

Berberis darwinii
Buxus sempervirens
Cornus spp.
Cotoneaster spp.
Escallonia spp.
Griselinia littoralis
Hippophae rhamnoides
Juniperus spp.
Laburnum anagyroides

Ligustrum spp.
Olearia spp.
Prunus laurocerasus
P. lusitanica
Rhododendron spp.
Rubus spectabilis
Symphoricarpos rivularis
Viburnum spp.

FOREST ENTERPRISE - Conifers

Abies alba
Larix decidua
L. eurolepis
L. leptolepis
Picea abies

P. sitchensis
Pinus contorta
P. sylvestris
Pseudotsuga menziesii
Tsuga heterophylla

Amenity planting

Abies grandis
A. procera
Araucaria araucana
Chamaecyparis lawsoniana

Pinus nigra
Sequoiadendron giganteum
Thuja plicata

Hardwoods-Amenity planting

Acer plantanoides
A. pseudoplatanus
Aesculus hippocastanum
Fagus sylvatica
Populus alba
P. nigra
P. x canescens

Populus serotina
P. tacamahaca
Prunus avium
Quercus borealis
Q. petraea
Q. robur

There are several fine plantations of estate woodland round Armadale and Dunvegan castles; in Portree, and at Lynedale (35) and Kilmarie (51). The first trees at Armadale were planted as a windbreak in 1785, with the main plantings 1815-30. Plantings began at Dunvegan around 1780, and continued there and in other 'big house' grounds through the 19th century.

The Forestry Commission had planted trees in Skye since the 1930s, greatly increasing their area in the 1970s but not so active since becoming 'Forest Enterprise'. The oldest plantings, which showed what could be done in spite of the wind, were at Glen Brittle and Eynort (now cut). Planting was extended to large areas of moorland in central Skye in the 1980s, by 'commercial' (tax avoidance) forestry - most of it consists of Sitka spruce.

APPENDIX III — **Botanists in vice-county 104, mainly Skye or Rum**

(Ra = Raasay, So = Soay, Ru = Rum, E = Eigg, M = Muck, C = Canna, Sc = Scalpay)

1700-1900

Martin Martin	ca. 1690	J.H. Balfour	1841
Sir J. Macpherson	1764	Rev. J. Fraser	1864
Rev. Dr. J. Walker	1764	Rev. H.E. Fox	1868
Prof. Dr. John Hope	1768 ?	Prof. M.A. Lawson	1868
J. Robertson	1768	E.F. Linton	1884
Rev. John Lightfoot	1772	W.R. Linton	1884
George Don	1794	Symington Grieve	1886
John Mackay	1794	H.C. Hart	1887
Dr. Graham Home	1825	W.R. & E.F. Linton	1888
John Home	1825	G.C. Druce	1898
W.G. Stables	1832	P. Ewing	1899
Prof. C.C. Babington	1841	Dr. S.M. McVicar	1899

1900-1950

G.B. Neilson	1902	U.K. Duncan	1940
A. Wallis	1909	W.A. Sledge	1947
P. Ewing	1911	U.A. Vincent	1947
Prof. J.W. Heslop Harrison	1934 (*)	S.Y. Barkley	1948 (So)
H.M. Mountfort	1934	F.R. Browning	1948
R.M. Adam	1938	J.K. Morton	1940s
J.E. Lousley	1938	(*) Ra, Ru, E, M, C, Sc, So	

1950-2003

P. Adam	J.H. Fremlin	J.R. Palmer
H.H. Birks	J.C. Gardiner	J.H. Penson
H.J.B. Birks	G.H. Halliday	D.A. Ratcliffe
R.A. Boniface	G. Hendry	J.E. Raven
A. Brewis	G. Hodson	J.G. Roger
F.R. Browning	G. Johnson	G.P. Rothero

J.C. Brownlie	A.G. Kenneth	J.B. Russell
J.W. Clark	J.B. Kenworthy (Ra)	A.A.P. Slack
R.W.M. Corner	R. Mackechnie	A.H. Sommerville
M. Coulson	P. Marsh	O.M. Stewart
K. Creighton (Ra)	D.N. McVean	A.McG. Stirling
N. Dennis	B.A. Miles	R.I. Sworder
U.K. Duncan	H. Milne-Redhead	M.G.V. Thompson (Ra)
R.E.C. Ferreira	H.M.J. Monie	E.C. Wallace
G.D. Field	C.W. Muirhead	M. McCallum Webster
R.S.R. Fitter	C.N. Page	J.E. Young

Botanists in the rest of v.c. 104 include:

Mrs E. Anderson (C)	M. Burnip (E)	C.A. Sinker (So)
Mrs F.M. Aungier (C)	J. Chester (E)	D.H.N. Spence (E)
J. Bevan	P.H. Davis (E)	Mrs E.V. Woods (So)
P.F. Braithwaite	R.H. Dobson (M)	P. Wormell (Ru)
S.J. Bungard (Ra)	R.M. Dobson (M)	

Botanists living on Skye, 1950 – today

I. Boag	M. MacDonald	C.W. Murray
P. Burke	C. Mackenzie	R.M. Murray
L. Cranna	M. Maclean	N.J. Murray
A. Currie	M. Macleod (Ra)	B. Philp
Mrs A. Gordon	E. McInnes	F. Reid
S. Gordon	A.C. Methven	D. Robertson
M. Gregory	M. Milne	E. Vickers
M. Henriksen	C. Murray	A. Young

Also BSBI Field Meetings in 1958, 1966, 1969, 1978, 1989; Wild Flower Society, 1975.

Edinburgh NHS	Lochaber NHS	Inverness Botany Group

INDEX OF LATIN & COMMON NAMES

Arabis hirsuta, 51
Arabis petraea, 51
Arctic Bearberry, 53
Arctic Mouse-ear, 43
Arctic Sandwort, 42
Arctium minus, 79
Arctostaphylos alpinus, 53
Arctostaphylos uva-ursi, 53
Arenaria balearica, 42
Arenaria norvegica, 42
Arenaria serpyllifolia, 42
Armeria maritima, 47
Arrhenatherum elatius, 99
Arrowgrass, 88
Artemisia vulgaris, 86
Ash, 73
Aspen, 48
Asphodel, 103
**Asplenium
    adiantum-nigrum**, 35
Asplenium marinum, 35
Asplenium ruta-muraria, 36
Asplenium scolopendrium. See
    Phyllitis scolopendrium
Asplenium septentrionale,
    36
Asplenium trichomanes, 35
Asplenium viride, 35
*Asplenium. trichomanes-
    ramosum. See* Asplenium
    viride
Aster tripolium, 86
Astrantia, 66
Astrantia major, 66
Athyrium filix-femina, 36
Atriplex glabriuscula, 41
Atriplex hastata. See Atriplex
    prostata
Atriplex laciniata, 41
Atriplex patula, 41
Atriplex praecox, 41

Atriplex prostrata, 41
Autumn Gentian, 68
Autumn Hawkbit, 80
Autumnal Water-starwort, 72
Avens, 58
Awlwort, 52

Babington's Orache, 41
Baldellia ranunculoides, 87
Barbarea intermedia, 50
Barbarea vulgaris, 50
Barren Brome, 101
Barren Fescue, 97
Barren Strawberry, 58
Bay Willow, 48
Beaked Tasselweed, 89
Bearberry, 53
Bearded Couch, 102
Bedstraw, 77
Beech Fern, 35
Bell Heather, 53
Bellis perennis, 86
Berula erecta, 67
Beta vulgaris, 42
Betonica officinalis. See Stachys
    officinalis
Betony, 70
Betula pendula, 40
Betula pubescens, 41
Bifid Hemp-nettle, 70
Bilberry, 54
Bindweed, 68
Birch, 40
Bird Cherry, 61
Bird's-foot-trefoil, 61
Bird's-nest Orchid, 104
Bishopweed, 67
Bistort, 45
Biting Stonecrop, 55
Bittersweet, 68
Bitter-vetch, 62

Water Mint, 72
Water Sedge, 95
Water-cress, 50
Water-lily, 37
Water-milfoil, 63
Water-pepper, 45
Water-plantain, 87
Water-starwort, 72
Wavy Bitter-cress, 50
Wavy Hair-grass, 100
Welsh Poppy, 39
Western Gorse, 63
Whin, 63
White Beak-sedge, 93
White Campion, 45
White Clover, 62
White Ramping-fumitory, 39
White Sedge, 94
White Water-lily, 37
White Willow, 48
Whitebeam, 61
Whitlowgrass, 51
Whorled Caraway, 67
Whorled Mint, 71
Whorl-grass, 98
Whortle-leaved Willow, 49
Wild Angelica, 67
Wild Carrot, 68
Wild Garlic, 103
Wild Onion, 104
Wild Pansy, 48
Wild Radish, 52
Wild Strawberry, 58
Wild Thyme, 71
Willow, 48
Willow Spiraea, 56
Willowherb, 63
Wilson's Filmy-fern, 34
Winter-cress, 50
Wintergreen, 54
Wood Anemone, 38

Wood Avens, 59
Wood Bitter-vetch, 62
Wood Dock, 47
Wood Forget-me-not, 69
Wood Horsetail, 33
Wood Meadow-grass, 98
Wood Sage, 71
Wood Small-reed, 101
Wood Speedwell, 74
Wood Vetch, 62
Woodruff, 78
Wood-rush, 90
Wood-sedge, 94
Woodsia alpina, 36
Wood-sorrel, 66
Woundwort, 70
Wych Elm, 40

Yarrow, 86
Yellow Corydalis, 39
Yellow Flag, 104
Yellow Iris, 104
Yellow Oat-grass, 99
Yellow Pimpernel, 55
Yellow Rattle, 76
Yellow Saxifrage, 56
Yellow Water-lily, 38
Yellow-sedge, 95
Yorkshire-fog, 100

Zigzag Clover, 63
Zostera marina, 89